THE ASH AND THE BEECH

Richard Mabey is the father figure of modern nature writing in the UK. Since 1972 he has written some 40 influential books, including the prize-winning *Nature Cure*, *Gilbert White: a Biography*, and *Flora Britannica*. He is a Fellow of the Royal Society of Literature and Vice-President of the Open Spaces Society. He spent the first half of his life amongst the Chiltern beechwoods, and now lives in Norfolk in a house surrounded by ash trees.

'This is the book of range and ambitions that his many admirers hoped he would write. Refreshing, droll, politically alert, occasionally self-mocking, he has the enviable ability both to write historical overview and also to slip into the woods like a dryad, bringing us back to the trees themselves, their colours and lights and textures'
Guardian

'Like the woodlands itself, *Beechcombings* operates on many levels . . . Busting out is a leaf-storm of philosophical musings, journeys of mind and body, reflections and anecdotes that imprint the tree on human culture'
Sunday Times

'Few writers can interweave personal insights, intellectual history and botany to create such a mesmerising and evocative account of man's relationship to nature'
House & Garden

'Mabey is a superb naturalist, but this is far more than natural history: it's history, art, memoir and occasionally, poetry'
Susannah Herbert, *Sunday Times*

'A loving tribute to the beech tree in history, social economy and the landscape is Richard Mabey's springboard for a wider story celebrating the character of tree and our responses to them'
Saga

'Once again, he has brought out something profound, special and startling – something that makes the rest of us wonder if we've ever actually been to the countryside, still less ever noticed anything special out there. Mabey's writing has roots'
BBC Wildlife Magazine

RICHARD MABEY

The Ash and the Beech

The Drama of Woodland Change

VINTAGE BOOKS
London

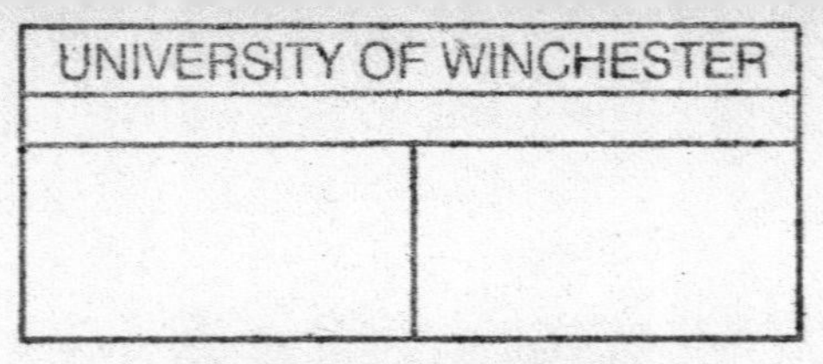

Published by Vintage 2013

2 4 6 8 10 9 7 5 3 1

First published in Great Britain as *Beechcombings* in 2007 by Chatto & Windus

First published in paperback in Great Britain in 2008 by Vintage
Random House, 20 Vauxhall Bridge Road,
London SW1V 2SA

www.vintage-books.co.uk

Addresses for companies within The Random House Group Limited can be found at: www.randomhouse.co.uk/offices.htm

The Random House Group Limited Reg. No. 954009

A CIP catalogue record for this book is available from the British Library

ISBN 9780099587231

The Random House Group Limited supports the Forest Stewardship Council® (FSC®), the leading international forest-certification organisation. Our books carrying the FSC label are printed on FSC®-certified paper. FSC is the only forest-certification scheme supported by the leading environmental organisations, including Greenpeace. Our paper procurement policy can be found at:
www.randomhouse.co.uk/environment

Printed and bound by Clays Ltd, St Ives PLC

Contents

For Bob and Libby

Foreword

These days I gaze out of my study with a mite of foreboding, waiting for a premature and maybe terminal autumn. Rooted in the bank of the ancient pond beyond the window are two multi-trunked ashes, airy, sprawling trees which together form a canopy stretching 20 metres across. They're fine at the moment, but just a few miles further north is the wood (its name, Ashwellthorpe, now seems an eerie black joke) where ash die-back first appeared in the wild. All winter the gales have been blowing Chalara spores south and west, and it's almost inevitable that the fungus will reach our garden. If these two trees are smitten, it will change the whole feel of our home patch. They have a spaciousness that ashes rarely have the chance to reach on narrow hedgebanks or in the tight ranks of woods. They're amphitheatres for bird flocks, vast and dramatic weather-vanes. Ash branches are elastic, and when they flail in the wind it is as if waves of wood are breaking across the garden. Losing them would mean not just a transformation of our view, but an unsettling shift in our sense of what constitutes a landscape, and what it contributes to a sense of home and security.

Much of Britain waits in a similar mood, wondering what the country will look like without our third commonest tree. Ash doesn't have the craggy grandeur of oak or the voluptuous grace of beech. It's short lived, usually collapsing at about 200 years, unless it's been coppiced or pollarded. Its pale trunks and filigree leaves, and a habit of regenerating in dense colonies, make it an often unnoticed choral background in woods, a visual hum behind the strong *timbres* of the big trees. But it's this quality that we love in it, that quiet, pale, graceful, background presence. Woods will, for a while, look emptied of *depth* if the disease hits badly.

And in hedgerows they make up a tenth of all mature trees. Most of the older individuals are pollards, low-slung and often cloaked with dense second-storey thickets of ivy, so these too are easily passed by, unremarked. But we will notice their absence, if and when they go.

But catastrophising (entirely understandable in the wake of Dutch Elm Disease) isn't a helpful response to threats to trees, and our anxious concern for them is easily trumped by our ignorance of their survival skills and community life. So this first spring after the Chalara's arrival in the UK I make a dispassionate surveyor's tour of the garden, trying to imagine what it will really be like if these two great sheaves of wood, and half a dozen younger trees, succumb. And, close to, the portents don't look so bad. Our ashes are surrounded (as they are in many places) by thorn trees, burgeoning self-sown oaklings, suckering wild cherries. In ten years' time the gaps they leave will have closed up, and the ashes will be metamorphosing into complex catacombs of decaying wood, full of beetles and woodpecker probings.

We have a cultural block against looking at trees like this, as dynamic and evolving vegetation. We want them to stay exactly as and where they are, and don't entirely believe either in their powers of self-regeneration or their afterlife. In an unstable world they've become monuments to security, emblems of peacefulness. We hug them, plant them as civic gestures and acts of reparation, give them pet names. When this cosy relationship is turned upside down – as it was, for instance during the Great Storm of October 1987 – we're shipwrecked ourselves, wondering if we've been bad guardians, not protected them enough. 'Trees are at great danger from nature' warned the Tree Council after the '87 storm – in an extraordinary solecism which seemed to place the arboreal republic entirely inside the kingdom of man. Very rarely do we ask whether we might have mothered them too much.

When Chalara struck the UK in 2012 it was clearly, in part, a breakdown in proper stewardship. The general public (and a good number of landowners) had never heard of the disease, but woodland ecologists and commercial foresters had been nervously tracking its inexorable westward march across Europe since the mid-1990s. Some urged the government to impose greater restrictions on the import of ash saplings, but most had few ideas about how to interpret or react to it. That is not surprising. The fungus, now dubbed *Chalara fraxinea*, is biologically mysterious, an entirely new organism of uncertain origins, which probably evolved in eastern Asia, where it appears to be harmless to native ash species. Its ancestor is a benign leaf fungus called *Hymenoscyphus albidus*, widespread, and native even in the UK. But at some recent date, this threw up a mutant, *H. pseodoalbidus*, with slight genetic differences but a terrible new virulence. The windblown spores infect ash foliage in spring, turning the leaf-tips brown. The fungal 'roots' (hyphae) spread through the leaf stalks into the branches and trunk, blocking off the tree's water supply. Typical diamond-shaped lesions appear on the trunk, and the leaves turn brown and wilt. Young trees can die within a year, but older ones appear able to survive for much longer. The fungus forms its spores in the leaf litter in summer, and these are dispersed in the wind over the following months. This is effective at spreading the disease over relatively short distances, but wind dispersal is limited by the fact that the spore can only survive in the air for a few days. In Norway Chalara has moved 20–30 km a year.

The first European cases were recorded in Poland in 1992. It had reached Lithuania by 1994, and then moved west and north, arriving in Italy, France and the Netherlands between 2007 and 2010. In Denmark the susceptibility of trees proved to be almost total, with not much more than 1 per cent left alive since the disease first arrived there in 2003.

It was this remorseless, epidemic contagiousness that caused such alarm and confusion when Chalara was spotted here, first on nursery saplings imported from Holland, then on wild trees which it can only have reached on the wind. Fantastical statistics were banded about in the media – that 30 per cent of all Britain's trees were ashes, and that with a host of other tree diseases already established here, we were facing a dead and denuded landscape, like the Somme after the Great War. In fact, Britain's ashes make up a little over 5 per cent of our tree cover, and are highly diverse genetically. The consequences of this variability, in terms of disease susceptibility, is already making itself shown in Poland, the first country to be hit. Between 10 and 25 per cent of Polish ashes are showing some level of natural immunity. In closely monitored populations in Lithuania, 10 per cent of trees have survived infection for 8 years and appear to be able to pass the resistance on to their offspring.

Natural resistance is likely to be the best hope for the survival of a core population of UK ashes. Isolated from the continent for nearly 8000 years, our trees may be more genetically diverse than those in Poland. For example, ashes that thrive in the sparse clitter of Yorkshire limestone are quite distinct from the tall poles that grow in damp East Anglian loams, and neither will survive if transplanted to the other habitat. Many ashes have male and female branches (and therefore flowers) on the same tree, so the potential for complex cross-pollination and extreme genetic variation is high.

It's a relief that the government has for once listened to its scientists and based its response on giving time and space for natural resistance to appear, and then capitalising on it, if need be, with cross-breeding programmes. Sanitation felling, which was talked about in the first wave of panic, would have been worse than useless, doing the disease's work for it, eliminating potentially resistant trees, and throwing more dormant spores into circulation.

But this laissez-faire approach isn't much liked. The public cry is for 'something to be done', for the excoriation of scapegoats in what is as much a natural event as a bureaucratic disaster, for raising the barricades, conjuring up a new woodland estate for the next generation. How have we come to regard trees like this? Human products, or worse, dependent arboreal children, only capable of appearing if we artificially inseminate the ground. Vulnerable to abuse from outside agencies ('nature' or nasty foreign organisms), but never from ourselves, and best put out of their misery if they become ill or old.

Understanding how these stereotypes and attitudes originated, and what perpetuates them today is crucial if we are to make a proper cultural response to and an accommodation with ash die-back, and with the many other diseases that are likely to affect our trees in the decades to come. *The Ash and the Beech: The Drama of Woodland Change* is a reflection on these cultural framings, a brief history of the narratives we've constructed about trees over the past thousand years, to make them accessible, useful, comprehensible and obedient to us. It's about the ideal forms created by artists, the explanatory dogmas of forest scientists, the fashionable plans of landscape designers. About trees as status symbols, political icons, emblems of reparation, and as investments, legacies, heritable goods. About the demonisation of trees that 'go wrong', become twisted, senile, decayed. About our new discovery of their crucial importance as regulators of the climate.

This would be a daunting task for the whole tree community, and in the book's first edition I chose to approach it through one species, the beech. The book is not exclusively about this tree – there is much on its relations with the ash and the oak – but the species is always there, as a kind of lens. There are personal reasons for this. The beech has been the key tree in my own life. I was born and grew up in the beechwoods of the Chilterns, and ran wild in them as a child. I seem always to have had beeches at hand as companions, or at least a kind of clock-face. One grew, quite unbidden, in the garden of the house where I lived most of my life. I had a table and chair under it for summer afternoons, an alfresco study. I'm living in East Anglia now, outside the tree's current natural range, and our garden is dominated by ash and cherry. But a planted beech – nestled among the municipally-protected chestnuts and oaks in a neighbours' garden – is still the first tree I see on waking in the morning. It wavers and swells if I move my head on the pillow – a refractive trick of the old glass in the windows, though beeches, sinuous and shape-shifting creatures, seem able to do this anyway. It's been the species which has framed my own view of trees – from feral childhood to rather studious middle-age. In the end it became part of the furniture and fittings of a piece of personal real estate, a wood of my own, which I bought and ran as a community project in the 1970s and 80s.

But the beech is also a fascinatingly awkward tree. It subverts clichés, refuses to behave as an upstanding parable of sturdiness and deep roots. It's unpredictable, possessive, prone to catastrophes – just as, unexpectedly, the ash has become. Our history of trying to make sense of the beech's contrariness is a reminder of just how far we will go to make nature over in our own image.

What follows is a set of discursive essays around some key episodes in the history of trees in Europe. The arrival of trees in Britain after the Ice Age; their early use as fuel wood and building timber; the apparent timber crisis during the naval wars of the 17th and 18th century, and the consequent invention of the plantation. Then the development of deliberate tree planting as a way of establishing status and legacy; the exploration of natural beauty through trees in the 18th century – for me, the pivotal chapter in the book; the attack on the last wild, wooded commons in the 19th century, and how they were saved by new ideas about the value of naturalness; the 20th century's flirtation with tree-spirits, ecologists' analogous attempts to explain the mysteries of tree succession, and the mythical idea of a 'climax forest' – all of which were to turned upside down during the storm of 1987.

The lessons of the Great Storm, explored at length in Chapter Seven, ought to inform how we respond to ash die-back, practically and culturally, and there should be no rushed or aggressive action. There was more damage caused to our woods by reckless clearing-up after the storm, than by the wind itself, and living trees, millions of seedlings and even the topsoil was often swept away by bulldozers, responding to political pressure and the public distaste for what appeared to be 'untidiness'. The contrast between the miserable replanting in these areas, and the spectacular re-growth in areas left completely alone has been a lesson which has still not been widely absorbed.

That favourite GP's phrase 'watchful waiting' is also appropriate. There is still much to learn about Chalara – about, for instance, its speed of spread and which ages of trees are most susceptible. The detection – and protection – of trees that seem to be resistant must be highest priority. So, wherever issues of safety aren't important, should the preservation of larger trees which succumb. A 'dead' tree is still a tree, and provides a rich habitat for bird, insects, fungi and mosses.

Replanting, especially with ash, should not be an option, until seed from local, resistant trees is available. Ashes from any other source (especially intensive nurseries both at home and abroad) are likely to quickly succumb, and may even introduce new strains of the disease. Tree species which naturally colonise areas where ashes have died are likely to be, by definition, those most suited to the site. Even where these are non-native (e.g. sweet chestnut, turkey oak, sycamore) they should be welcomed, with tolerance as well as vigilance.

Sycamore, of course, is currently demonised as an 'invasive alien', introduced sometime in the late Middle Ages (though it is quite possibly indigenous, given to erratic and untypical behaviour for a native because of its own fungal affliction, black spot). But we should do our best to make an accommodation with it, as perhaps the best natural coloniser of bare patches that is currently available. It can't host many of the insects that have co-evolved with ash over thousands of year, but it will be partial refuge for the lichens which are ashes' outstanding familiars, and restore at least an ambience of woodiness. Climate change is making the categories of native and non-native increasingly fuzzy, and we may find ourselves grateful for some immigrant biodiversity.

Above all, the lesson of the storm was that catastrophes – be they disease, climatic trauma, insect predation – are entirely natural events in the lives of trees and woods. They respond, adapt, regroup. What emerges in their recovery stage may not be the same as before, but it will always be a vital, dynamic, arboreal community. The same process will happen with ash, perhaps more quickly than we think.

Norfolk 2013

Chapter One: 'The Lowest Trees have Tops'

'The lowest trees have top, the ant her gall.
The flie her spleene, the little sparke his heate.'

Attributed to Sir Edward Dyer (1543–1607)

I

I CAN NEVER MAKE it out from a distance. I must know every twist of its knotted trunk and serpentine branches, but a couple of hundred metres away it's just part of a general woodiness, a green blur. It's called the Queen Beech, but in reality it's just a commoner, one of a grove of ancient beeches that have grown on the waste above Berkhamsted since at least the Iron Age. Trees, even the teeming mosaics of individual woods, always vanish in the long view. Move closer and different images, different perspectives, swim into sight. At a hundred metres I can see the Queen as a separate tree, a member of a species, a manageable object. But at thirty it's a wild individual, its immense and snaky branches beyond any accounting or control. At a dozen I'm inside it, and it's scarcely a tree at all. It's a catacomb of labile wood, a veil of translucent leaves exchanging breaths with me. Is any of these images truer than the rest? Is there such a thing as a real tree, beyond the images we make of them as lovers of views or curious naturalists or harvesters of timber?

I've been coming to this wood since I was a boy. It's called 'Frithsden Beeches' – 'a beechwood in a wooded valley'. It's an old and cryptic place, and its long history is engrained in the trees. They were lopped for fuel by early peasants, looted by the Normans and almost lost during the enclosures in the 19th century. Now, as unseasonal gales rip across the Chiltern plateau, more and more of them are entering a new phase of existence, as horizontal trees. At every stage in my life they've had a different meaning. When I was young they were my benchmarks and touchstones, and I found a kind of security in their ancient quirkiness. I gave them shamelessly anthropomorphic names. Falstaff – low-slung, bulbous, cankery, undeniably jovial. The Organ-pipes, topped with an immense Gothic spray of vaulted branches. The Praying Beech, with two branch stubs fused in the form of a pair of clasped hands. The nameless tree with a muddled frieze of ancient branch-stubs that looked like nothing so much as an X-ray of the human digestive system. Even when they were blown to the ground, they still looked lucky: elbows in, paunches cushioned in the mud, roots like flagstaffs – or like saplings to come.

But the Queen has never looked like coming down in the gales. Hunched at the very edge of the wood, just where the Beeches abut onto the open common, it's an antic and indomitable matriarch. I gaze at it, for the umpteenth time. It seems elephantine, an impossible mass for a living thing. It is, I guess, between 350 and 400 years old: two centuries of being repeatedly beheaded for firewood, two more as a picturesque monument. It grew up in the open, unrestricted by other trees, and its long low branches trail out like the arms of a giant squid. Its trunk is vegetable hide, a mass of burrs, bosses, wounds, flutings, folds of scar tissue congealed around the points where the branches were lopped. One storey up there are mosquito pools in forks, old woodpecker holes, generations of graffiti. Some of the scratchings are in implausible positions: the higher you carve your message, the code reads, the more impressive your feelings. With my binoculars I can just make out some of the inscriptions. The names and homesick addresses of American servicemen stationed nearby during the Second World War. The linked pledges of sweethearts from the outbreak of the First. The copperplate initials of Victorian schoolboys, now stretched beyond deciphering. The letters 'S.A.' many times. A heart. A rose. Not really tree-abuse, as it's so often reckoned, nor always a compulsion to leave one's mark on the world. More, I think, the result of the world's leaving a mark on you. No one encounters trees like this without some kind of conversation taking place, an exchange that deserves a memento. Beech-scribbling goes back to classical times, and has its own Latin epigram: *Crescent illae, crescit amores.* 'As these letters grow so will our love.'

I once tagged along behind a party of forestry professionals on a tour of the Beeches. They were a gaggle of estate managers and big landowners, out to check each other's woodland growth against their own, or against some abstract ideal of tree productivity. They were outraged that this collection of 'mutilated freaks' had been given living space for so long. 'They're rubbish,' growled a local major, whom I'd last seen in his own wood, shovelling up piles of men's magazines as if they were leaf-mould, 'an insult to the forester's craft.' There was talk about the dangers to walkers, of crippling insurance claims. The consensus was that the whole lot should be summarily felled and replaced with 'proper' trees. The owners of the Beeches (the National Trust), remembering that the wood had been saved from destruction during the 19th-century enclosures by an epic local uprising, put up notices instead, their own gesture towards inscription: 'These very old pollarded trees and associated deadwood in this area are being managed for their nature conservation and historic interest. They are liable to shed branches and the public is advised to keep to the waymarked rights of way.' The public resists the advice, feeling much as the ancients did, that as these trees grow so does their love.

In those days, the Queen Beech was my party piece. I led people through the mazy thickets of the common and unfurled it like a bunch of flowers from a conjuror's hat. *That* is how tough nature is, I think I was saying. That is what a tree can become, beyond our conceptions of perfection and usefulness. But that was as partial a view as the major's, a romantic's hope that nature might show us the way. These days I stand more pensively at the Queen's foot, earth-bound, dwarfed, gazing up. Pondering perspectives. What does a tree seem like to a creature perched on the top, looking down on the immense supporting mantle that we groundlings call the canopy? What is its own version of the agenda of survival, its own attitudes towards shape, productivity, decay? I remembered the transformed view of the world we had as children, hanging upside down from the branch of a tree. All the details the same, but in a different order, a different hierarchy.

The Czech poet Miroslav Holub's poem 'Brief Reflection on Cats growing in Trees' imagines how trees might be interpreted from a mole's-eye view. The moles emerged at different times to report on the way things were above ground. The first saw a bird on a branch, and reported that 'birds grow on trees'. The second, surfacing later, saw a cat, and concluded that cats were the true fruit. The confusion worried the top mole, so he went to see for himself:

By then it was all pitch-black

Both schools are mistaken the venerable mole declared.
Birds and cats are optical illusions produced
by the refraction of light. In fact, things above

Are the same as below, only the clay was less dense and
the upper roots of the trees were whispering something
but only a little.

Our views of trees are scarcely more inclusive than the moles'. We blink at them from our cultural burrows and see what we want to see. Models of beauty, wastes of space. Dangerous excrescences, and the dwelling places of gods. Workhorses, ornaments, investments. Source of the crown of thorns and the olive branch. Metaphors for the state, for the human body, for life itself. If the forest has always been an ambivalent idea in human consciousness – the contrary of civilisation, yet also its primary life-support system – so have the trees that comprise it. Trees, historically, have been a *challenge* to humankind. They are monumental, long-lived, stubborn, territorially ambitious. They don't fade into the background or live modestly on the peripheries. Trees occupy space. They insinuate themselves into every kind of environment. They are what dry land aspires to become. At repeated points in human history, it has seemed to be a case of them or us.

Mostly it has been us. The beginnings of agriculture, the founding of cities, the creation of energy systems based on fossil-fuels, were all made at the expense of trees. Collectively, the world's trees have been on a downward path for the past 5000 years. But we can never dismiss them entirely – not just because they produce much of the oxygen we breathe, but because they are the measure of us. They are our lost home, our epitome of nature, one of the benchmarks by which we judge, for better or worse, our standing as a species. To be without trees would, in the most literal way, to be without our roots.

So we try to reach some kind of accommodation. We've cast trees in roles, negotiated with them. In hunter-gathering societies, it was often thought proper and necessary to placate trees' spirits when they were cut down. Strip off some of the bark to set the spirits free. Slip a wad of herbs into the soil as a votive offering. Later, in the temperate zone at least, there was a kind of secular equivalent to these rites of arbitration in practices which took a continuous crop from trees without killing them. Nature itself would do the renewing. But for the last three centuries, we've increasingly appropriated for ourselves the role of regenerators. We've deployed trees across the landscape as if they were incapable of doing it themselves. To satisfy our timber needs we plant and harvest them like arable crops. We raise them as screens for eyesores and dignifiers of developments. Children are encouraged to plant and coddle them to nurture their feelings for the natural world, as if they were pets. Only rarely are they seen for themselves, as autonomous, anciently evolved beings, quite able to sustain their own lives – and sustain ours into the bargain. The mutuality of our relationship is being forgotten. 'Trees give off carbon dioxide in the night and poison me' was the verdict of one London householder, so alarmed at the excesses of these intruders in her street that she notified the local environmental health officer.

Remarkably, trees still cover one-third of all dry land, in one form or another. They are the one kind of natural organism that most humans are rarely out of sight of for more than a few seconds. As plantation clones, desert dwarfs, virgin forest giants, they continue to be the primary engines for converting solar energy into the solid materials that all other organisms need for food and shelter. That fundamental chemical exchange – the light-activated reaction between carbon dioxide and water that's known as photosynthesis – also produces much of the planet's oxygen. Without trees, or something very like them, most of the Earth's present inhabitants, ourselves included, could not survive.

But what could that alternative have been? There was something inevitable about the evolution of the tree, this structure for defying gravity, for raising life above the ground – and above the confines of the present, too. Trees are the architectural climax of evolution, scaffolding for the rest of terrestrial life. Many widely different plant families – palms, club-mosses, buglosses – have produced them. If you were trying to devise a perfect plant form that had the same strength and durability as rock, it would be the trunk of a tree. In their maturity, not quite like any other living thing, they become increasingly complex, vast elaborations in three dimensions. As their branching becomes more intricate, so do the niches formed amongst the branches. A full-grown tree is a catacomb of reticulations, rot-holes, snags, fissures. Even the twigs develop architectural layers – flakes of bark, small bosses where smaller twigs have broken off, velvet sheens of moss. It's impossible to measure the area of a tree's surface exactly. It's what mathematicians call a 'fractal' quantity, one that increases indefinitely the closer you examine it. The American writer Annie Dillard's question to God in *Pilgrim at Tinker Creek* was: 'You want to make a forest, something to hold the soil, lock up solar energy, and give off oxygen. Wouldn't it have been simpler just to rough in a slab of chemicals, a green acre of goo?'

Some of the most majestic trees on earth, the giant coastal redwoods of California, actually change the structure of the ground. They're shape-shifters, securing their own and others' futures as generously as beavers damming a river. When the redwood's roots are drowned by high water they send up new roots vertically, which then sprout new lateral roots just below the level of the new silt, anchoring the tree and stabilising the ground. Along some of the coastal rivers, a thousand years of flooding have raised the level of the whole area by 9 metres – and the redwoods, every bit as old, have responded by developing multi-layered root-systems like inverted pagodas.

But the immense genetic intelligence of trees like these hasn't saved them from the floods of humanity. The coastal redwood's cousin, the 'Big Tree' of the Californian sierras (our 'Wellingtonia', but properly *Sequoiadendron giganteum*), was one of the most infamous victims of the opening-up of the American West. When the sequoias were first glimpsed by dirt-poor miners in 1852, they were looked on as wooden gold. They were incomprehensibly huge, 30 metres or more in girth, and rose beyond human sight. They might hold as much as 10,000 cubic metres of lumber, and even though it could take five men three weeks to fell a single tree, down they came. For a while the redwood groves in Yosemite became a kind of botanical amusement park. A two-lane bowling alley was built along the surface of half a trunk. The stump of one felled tree was made into a dance floor for the tourists, where, in the words of the entrepreneur who set up the show, 'thirty-two persons were engaged in dancing four sets of cotillion at one time, without suffering any inconvenience whatever'. But within ten years attitudes towards the redwoods changed. Those that remained standing began to be seen as a fundamental part of America's heritage, proof of the nation's ancient and sacred roots. In 1864, at the height of the Civil War, Abraham Lincoln signed a bill to create the world's first wilderness park, and granted the Big Trees to the state of California, 'for the benefit of the people, for their resort and recreation, to hold them inalienable for all time'.

Just a decade for official opinion to turn around – and just another hundred for it to go almost full circle, with North America's ancient forests coming under threat again. Across the globe (not least in

Britain, site of the earliest and most thorough forest clearance in Europe) we profess an understanding of the importance of trees while continuing to raze them into oblivion. The consequences have almost become clichés: erosion, flooding, the mass extinction of species, the disintegration of local cultures. Now we know that tree loss has contributed to climate change. Global warming is partly the result of recent deforestation, particularly the burning of forests, but chiefly of the extravagant release into the atmosphere of the carbon of prehistoric trees, locked up by nature under the ground. Yet there is an equivalence that should make us pause. The excess carbon dioxide in the atmosphere could be absorbed, for a couple of centuries at least, by about 10 million square kilometres of new tree-land – of the same order as the area we have destroyed globally since the start of the Industrial Age. After that, it would be slowly released again as the trees started to die and decay. But reforestation would buy us time, and the quibbles about its supposedly temporary benefits (only two hundred years!) are mostly sour grapes, from alternative energy suppliers. *All* new trees are important now. But, at present, such a dedication of civilised land to wood is socially and politically unimaginable. So we make token gestures, plant a few trees in the school grounds, recycle an armful of newspapers. It all helps, but it is not the kind of heroic action that's needed.

The long pattern of our relations with trees begins to look familiar, the same shape as our paradoxical relations with nature itself: dependence and notional respect at first; then hubris, rejection, the struggle for dominance and control; then the regret for lost innocence, the return of passion, the pleading for forgiveness . . . A cynic might say this is exactly the love–hate see-saw that occurs in abusive relationships. But generalisations of this kind don't help. Both nature and humanity are too complex. Even 'tree' as the epitome of nature begs too many questions, invites a view of them as emblematic staffs of life, totem poles. Beyond the fundamental of a tree's life – the sun shines, the leaf breaks, makes oxygen, lays down mass – what happens next is both extravagant and particular. The tree's identity counts, not just, in the most basic of ways, to itself, but to all other beings that live with it. The coastal redwoods, cousins of the Big Trees of California, spring again from their stumps. The remains of 1,000-year-old giants cut down more than a century ago are surrounded by rings of their regenerated shoots more than 50 metres tall. No one expected the redwoods to be one of the few conifers that would coppice, and to be virtually immortal. Trees are individuals. Every species has its own habits, its own cargo of metaphor. Close-ups of the forest's green chaos help.

In the temperate zone eyes have traditionally focused on the oaks, a family of trees so useful and adaptable that they've provided, single-handed, most of the materials necessary for the development of technological cultures. Their bounty included handles for axes, bark for tanning leather, charcoal for fuelling iron-furnaces, boats for warring navies, galls for the first inks for the first natural history books.

The American arborist, William Bryant Logan, has made an audacious suggestion about the global role of the oaks. In *Oak: The Frame of Civilisation* he argues that it was specifically this huge family of beneficent trees that enabled humans to make the transition between hunter-gathering and settled cultivation. Archaeologists normally grant this role to the wild grasses of the Middle East, which made possible the development of agriculture, bread, permanent villages, and the division of labour – a way of living that was exported, for better or worse, across the planet. Logan's alternative is seductively argued. He has produced a map in which the distribution of early settled societies throughout the temperate zone appears exactly to coincide with the geographical spread of the 400-odd species of oak. He cites cultures in North America, ancient Mesopotamia, the highlands of Mexico, where a style of living midway between nomadic gathering and rooted agriculture was evident long before the advent of cereal farming. You could, I suppose, call it fixed foraging, the communal exploitation of a long-lived local resource. The resource was the oak tree, usually around in one form or another. Its first and most fundamental gift was the acorn – prolific, nutritious, cookable, storable. Acorns, Logan argues, were the world's first staple food. Then came the incomparable gift of oak-wood, tough, durable, cleavable. Oak planks made possible that ironically crucial stepping stone to civil society, the fence. Then they enabled migration, as the infrastructure of waterproof boats and of wooden walkways across the swamplands.

It's a beguiling case, and a more pleasing image of the natural transactions that gave birth to early civilisations than the bludgeoning march of agriculture. But like all Grand Theories, it overstates its case, and the uniqueness of oak as a cultural root and branch. In Kyrgyzstan, there are ancient semi-nomadic communities based around walnut and wild apple trees. In the Italian Appennines and many pockets of southern Europe, whole cultures were framed on the sweet chestnut, as a source of nut-flour, building timber and fuel wood. The early Turks built very serviceable ships from pine, elm and mulberry. Even in oak-proud southern England, it was beech not oak that provided most of the fuel for London, and for the iron- and glass-works of the Weald. There would always have been some kind of symbiosis between pre-industrial societies and trees. No other resource could provide such a range of food, fuel and raw materials. But trees are a multifarious tribe, and human ingenuity has usually been able to make something out of whichever happened to be at hand. It just happens that oaks were pretty well always at hand.

The ubiquitousness and usefulness of the oak have tended not only to obscure the value of other tree species, but to warp the image of the tree itself. Its qualities – strength, longevity, a kind of frontier spirit – have come to be seen as the quintessence of the 'good' tree. And sometimes, by association, as the quintessence of the places in which it grows. In 17th-century England famously, the oak became a symbol of national pride and naval aspiration, the spiritual – as well as material – source of the people's 'hearts of oak'. A case, perhaps, of not being able to see the tree for its wood.

The biographies of other trees frame other parts of civilised history. The small-leaved lime, once the commonest north European forest species, and the great wood-carvers' tree. The elm, favourite fodder species of early herdsmen, a building wood second only to oak, but an ancient victim of disease and symbol of death. The ash, pioneer coloniser of open ground, and abundant and basic rural wood source, for furniture, tools, firewood.

II

This book focuses on the beech and on its nogotiations with other trees, especially ash and oak. I've spent most of my life amongst beeches, and know them better than other trees. But there's an intriguing eccentricity about them, too. They don't conform to the image of the 'good' tree. They can appear in the archetypal form we imagine is correct for trees – a rounded bush of foliage on top of a straight pole – but most beeches come in odder shapes than this. They can be elongated, dwarfed, as bulky as oxen. They can have the look of sinuous strength, but be useless as building timber. They are elegant but also catastrophic, vulnerable to gale and drought. No one would ever sing about 'Hearts of Beech'. Some writers (myself included, in the past) have tried to magnify these differences to make a neat poetic contrast between the oak and the beech. The tree of robustness and the tree of fashion. The light-lover and the shade-bearer. The deep-rooter in clay and the frail haunter of thin soils. But the realities are more complex than that, the differences a matter of degree. All trees need light and some kind of stable base to root themselves in. All of them need mineral nutrients from the soil. And, as the Great Storm of 1987 showed, in its tipping-up of 15 million trees, almost all of them have much shallower root-systems for fulfilling these functions than was popularly imagined. The beech's roots are happier than the oak's in mineral-rich chalk and limestone soils – but their shallowness makes them more unstable on such sites. Both species flourish on thin, acid soils, but on well-drained sands their seedlings have trouble reaching nutrients, so regeneration can be poor. Checks and balances perhaps. Even their respective responses to shade don't suggest an absolute difference. Oak, which now seems to regenerate only in the open, did, until about 1910, grow quite happily under its own canopy (the change may be due to the arrival in 1906 of an American mildew that weakens its seedlings, especially when they're growing in high humidity inside woods). Beech, conventionally seen as the archetypal shade-giver and shade-bearer, needs some break-up in the canopy before its seedlings can grow. The only real generalisations that can be made about the two trees' relationship is that the beech's greater height and denser foliage give it the edge over the oak in the short term, but that its greater instability will eventually balance that out. Rather in the manner of a risk-taking actor, the beech can command an immense presence, but may at any moment fall flat on its face.

Out in the real world, trees break all the generalisations made about them. Beeches can survive hurricanes, but have a genetic tendency to split in two of their own accord. They are sensitive to drought, but supposedly die if their roots are waterlogged – until you discover them growing in a bog. Trying to make hard and fast rules about them is as futile as straightening a snake.

The family *Fagaceae,* which contains the families of both oak and beech, split off (from the marrows) about 90 million years ago. Some time later the beeches branched off from the oaks. There are just ten species in the genus – seven in Asia, two in North and Central America and one in Europe. The European species, *Fagus sylvatica,* is now widespread, and across many eastern parts of the continent is more dominant than the oaks. As a wild, indigenous tree it grows from southern Sweden to the northern reaches of Turkey. There are ancient beeches, draped with lichens, in the moist Atlantic air of the Pyrenean foothills, and bleached pollards in the hot mountains of Greece. Natural beechwoods stretch all down the central spine of Europe, through Normandy and the Alps and the German heartland into the Appennines. Beech grows with silver fir in the wood-pastures of the Czech Republic and Slovakia, and with dark yew and pale ash on the English chalk downs. There are native beechwoods in the very heart of the Mediterranean, on the island of Corsica. If you climb up from the parched coastal belt of rosemary and cistus, through belts of deciduous oaks, then of silver fir, you come to another layer of broad-leaved trees, the mountain beechwoods. They grow up to the edge of the snowline at 1,500 metres. I've walked them in early spring, before the leaves have opened. They're the epitome of European beechwoods, pale, gracious, airy, a touch monastic. But even without leaves they're shadowy places, the leaf-litter lit up here and there by meagre clumps of crocus and cyclamen. On flatter ground, and tucked in close to waterfalls, there are huge lopped trees, evidence that even in these remote heights, humans have been beavering.

But there are no wild, self-sprung beechwoods in the dry

Mediterranean lowlands, or in the cold north European hills. Planted beeches can survive in these regions, but may not be able to flower and seed sucessfully, because of drought or frost. Their sensitivity to weather is a check on all beeches, and keeps them on a narrow edge between triumph and collapse.

The roots underpin this continuing gamble. They've evolved as an adaptation to thin soils, and brace themselves across the surface, saturating the ground with secondary capillary roots. They draw most of the nutrients out of the topmost layers of soil, making it hard for other trees to establish themselves close by. In most sites they only reach 0.3 to 0.6 metres into the earth, in deep soils 1.2 metres at the most. In dry spells they turn upwards, towards the surface, to take the first advantage of rain. During long periods of high temperature that dry out the top layers of soil, beeches become dehydrated. Their outer twigs shrivel. Irrelevant branches may die. The thin soil round the roots becomes dusty and desiccated, and the whole exquisite architecture for making the best of any water that is available becomes as inadequate as a spider's web in a storm, a shallow lacework that can scarcely anchor the tree to the ground.

The beech's root-form helps determine not only where it can grow, but how it grows. Beeches in the open, or at the edges of woods, develop broad root-plates, mirrored by wide, low crowns. Inside woods, with less space and light, the trunks soar upwards but the roots can't always expand sideways to compensate. In the cramped conditions of plantations they're even less secure. Plantation forestry has spread the beech far beyond its natural 'comfort' zone (into northern Scotland, for example), but, packed in at high densities, the trees are especially vulnerable to hostile weather.

The trunk contributes to stabilising the tree. Flared buttresses can develop where root and trunk meet. The tree will try to keep its centre of gravity down, with long lower branches, but in the competitive shade of its neighbours will shed these early as it reaches up for light. In woods it can grow up to 40 metres tall, and the first 20 of these may be free of branches. This is the natural form of what are often called 'high forest' beeches. In the open, with room to spread, they'll branch much lower. Wordsworth described the 'Alfoxdon Beech' in Somerset, as 'throwing out arms that struck the soil, like those of the banyan-tree, and rose again. Two of the branches thus inserted themselves twice, which gave to each the appearance of a serpent moving along by gathering itself up in folds.' Beech always has this plasticity, responding to the opportunities of space and to disruptions of its growing pattern with extraordinary improvisations of form. As the oak tends towards angularity, a certain abruptness in the way its branches jut and turn, so the beech drifts towards sinuousness. Its branches curve upwards, the twigs emerge in sprays. Wounds are rounded off as if the wood were potter's clay. I've seen trees like immense candelabra, the outer branches driven into vertical growths by competition from surrounding trees. And I've once seen a bonsai beech, centuries old but no more than a metre tall, curling out of a crack in a vertical cliff. Somewhere, in the flat limestone karst of eastern Europe, I've no doubt, there is an entirely horizontal beech, creeping along in the damp shade of a crevice, as ashes do in the limestone pavements of northern England.

Up in the canopy, next year's buds appear in May, just after the opening leaves; by August the little spikes are already full of embryos. They hatch in late April and early May, and so rapid is the transformation of a beechwood by this effusion of sappy, luminous green, that I used to believe the leaves simply unwrapped, took on their full form in a matter of hours. But I've watched them more closely now. I've tied tapes round the twigs so that I can identify individual leaves. They do unwrap, in a sense. There is no organic growth, no cell-division in bud-burst. The burgeoning leaves are simply pumped up with water. But they take two days to inflate, from the first split in the brown husk of the bud to the full spread. A little longer than the hatching of a dragonfly, but with something of the same style. They are furled like flags, or fans, in the buds, with the folds parallel to the veins. As they open – flat against the light, like drying wings – they echo the splayed form of the clusters of twigs. A fringe of soft hairs, like cilia, or baby-down, persists for a while along their edges. The shape of each leaf, and of each bunch of leaves, replicates the form of the whole tree.

Beeches don't begin to produce flowers and seeds until they are at least 40 years old – sometimes as much as 80 years in dense, shaded woodland. The flowers are greenish and inconspicuous, the male a stalked tassel, the female a tuft. They appear together in May, when the tree is in full leaf. They're pollinated by the wind and should, in late summer, produce the small, spiky-shelled, three-sided nuts known as mast. But the beech is vulnerable at this crucial moment in its life, too. The flowers are highly susceptible to the cold. Their buds won't form at all if there's been poor sunshine the previous summer. As little as 1.5 degrees of spring frost will kill them altogether. Even if the flowers survive, mast-production is a chancy business, only generating heavy crops every three to six years (though this is quite sufficient to regenerate the tree). Between 1934 and 1944 in England there were no good mast years at all. Perhaps the all-or-nothing production of seed is a strategy by the beech to guarantee the survival of its next generation. Animals which feed on the mast – and the sprouting seedlings – would be less likely to eliminate an occasional huge fall than if, in the same numbers, they were feeding on smaller, regular crops.

But in good mast years, beech seedlings can be thick on the ground. If they survive to become mature trees, a stand of beech, they are uncompromisingly tribal. Other trees are rare amongst the dense-packed stems. The shadowing from their interleaved branches is hostile to almost all ground plants except mosses. Even beechlings can't prosper under their older generation's densest shade, and only begin to grow when the leaf canopy starts breaking up, from age or rough weather. But there are mysteries in even successful regeneration. How are the compact little fruits, with no way of catching the wind, moved about in a wood? Where is that delicate point where a beechling has just enough shade to keep its roots moist and its competitors subdued, but sufficient light to develop itself?

When you look at the beech's sensitivities it's a wonder that it has survived as a species. It ought to have been one of evolution's casualties, or confined to some mythical land where conditions were always warm and moist. A beech Atlantis. Yet almost as surprising is that any other kinds of tree have survived in its company, that beech hasn't become a kind of super-organism, shading out all potential usurpers of its ground-space, indefinitely monopolising the tree cover. In some places it does so, at least for a while, producing that intense grey architecture that, according to your point of view, is either awesome or monotonous. In such pure stands, beech can have the look of an ecological tyrant. It seems to contradict the rule that tree-growth enriches a habitat, generates niches for other species, encourages the diversity of life. The arrival of beech in a wood of other species changes the wood's character more than does any other tree. It can begin to resemble a woody desert, a forest reduced to the essentials of leaf and trunk and earth. Birds are few, insects invisible. The ground flora is not much more than a scatter of species, like bluebell and woodruff, that are shade-tolerant themselves – and then, in the gloom of high summer, a mysterious tribe of shadowy orchids. But this bleak image is part of our groundlings' tunnel vision. Beechwoods have occult lives of their own, in layers and time-frames we're not used to exploring. In old age especially, the tree can grow into a quite different being, a resilient, frugal survivor. An unlopped beech may live for about 200 to 250 years before it begins to die back, a pollard sometimes as much as twice that. In either form, an ageing tree may begin to lose the interior of its trunk. Branches, even the entire top of the tree, may be shed. Scar tissue grows round the broken ends, and around the lip of any hollows. Rot-holes appear, and other tree seedlings may take root in them. And into this increasingly craggy structure come whole new universes of insects, lichens, ferns, fungi. None of this downsizing is unnatural, or an indication of sinister deterioration. It is the way in which trees prolong their lives.

What does keep the beech in check is not ageing, but its vulnerability to extremes of weather. It's a predicament that may get worse. The tree's sensitivity to drought and storm make it a possible victim of global warming. The widely accepted scenarios for climate change in Britain all have implications for the future of beech. The 'Low' model envisages a 0.5 degree increase in mean winter temperatures in the 2020s, and a 0.9 increase in the 2050s. In the 'High' scenario the figures are 1.4 and 2.3 degrees respectively. Years with severe summer droughts – rainfall less than half of the long-term average – will increase tenfold by the 2050s, from the present level of one per century. Wrecking gales will become more frequent.

One severe drought every ten years is hardly chronic stress. But it's likely to erode existing populations of beech in drier areas such as East Anglia, Kent, southern Hampshire and the Wye Valley. Foresters are already noting a deterioration in what they call, a little portentously, 'the condition of beech'. In the years following recent hot dry summers, trees across southern England have developed a set of related symptoms. They go thin on top, and their 'crown density' declines by as much as a quarter. This is a visual judgement made on the basis of the 'transparency' or lack of leafiness in the crown, which is caused by the die-back of smaller branches and the production of fewer leaves. The leaves themselves become smaller, and may show signs of rolling up at the margins, a response to limit water loss. They may lose their usual colour and be shed earlier for the same reason. Beeches stressed by drought can become more susceptible to fungal diseases and predation by insects or bark-nibbling squirrels. In extreme cases they die, as happened widely after the drought summers of 1976 and 1994. In the Chilterns, with its large concentration of beeches, foresters were predicting that the tree would be extinct on chalk soils by the end of the 20th century. But it didn't happen. The tree continues to prosper, though not always in the fulsome forms preferred by commercial growers. A beech which is retrenching in response to drought is still a viable tree.

With beech trees, their problems with future droughts may be partly balanced by the benefit they get from rising temperatures. When wild beech arrived in this country after the Ice Age, it failed to reach areas such as Cornwall where it ought, theoretically, to have been entirely happy. A future reduction in the number of flower-killing frosts means better mast-production, and the chance that the tree could become self-propagating far beyond its current range, and at higher altitudes too.

But we're always inclined to choose the gloomiest options when imagining trees' possible futures, casting them in the role of vulnerable children, or arboreal pets, or, sometimes, scapegoats for our own mistakes. Short-term problems with weather or disease are magnified into unfolding catastrophes, tree panics. In the 1980s, 'stag-heading' (a budgeting move by trees when they need to economise on water) was widely prophesied to be a prelude to 'the End of England's Oaks'. When it plainly wasn't, attention switched to oak 'knopper galls', the deformed and sterile acorns, rather like pleated medieval hats, which result from an invasion by a continental wasp-grub, *Andricus quercuscalicis*, and which were seen as likely permanently to emasculate our national tree's reproductive powers. Twenty years on, the wasp is still about, and still produces a few galls on a few trees, of no significance whatsoever amongst the myriads of fertile acorns. In the new century has come the spectre of 'sudden oak death', which has so far only affected a handful of garden shrubs quite unrelated to the oak.

Disease in trees can, of course, lead to disaster, especially when it originates from introduced organisms to which native species haven't evolved any defences. Ash is now threatened by an entirely new fungal organism that may have evolved in the near east. A canker-producing fungus from Japan decimated American sweet chestnuts in the 20th century. Dutch elm disease, which is now known from fossil evidence to have been around in Britain since prehistoric times, virtually wiped out the English elm and reduced most small-leafed elms to the status of shrubs when a new and virulent strain arrived in wood imported from Canada. As I write, there are headlines describing a new threat to horse chestnuts, described in the popular press as 'a triple whammy' of drought, trunk canker and attack by leaf-miner moth caterpillars. Conker trees have been dying since the 1960s from fungal invasions of their sapwood, and more are likely to be prematurely felled because they are believed to be dangerous. As for beeches, they too have had their share of afflictions, from plagues of beech aphids to wet-rots – all of them more serious after summers of drought.

Oliver Rackham tells a cautionary story about the ancient beeches in the Pindos Mountains in north-west Greece. Pindos is a remote area, far from any industrial centres. But in the 1980s the beeches showed all the signs of another fashionable affliction – *Waldersterben,* forest death from acid rain. The symptoms were the familiar ones of stress: die-back at the tops of trees, leaf-yellowing, early leaf fall. But many of the huge trees were then more than 300 years old. They were covered in luxuriant lichens, a sure sign of unpolluted air. And their narrow annual rings showed they had been in this state of retrenchment for centuries. Reduced vitality was the reason for their survival.

It's sensible to be vigilant with all tree species, especially at a time of rapid environmental change. But ageing, coexistence with parasites, retrenchment, are not 'diseases'. They are ancient arrangements trees have made with their environments. More worrying would be trees which appeared entirely 'healthy', unnibbled by insects, abandoned by fungi. What would that say about the toxicity of their surroundings?

Many of our concerns about the fragility of trees are exaggerated because we tend to perceive them as human artefacts – and sometimes as humans themselves.There's a kind of anthropomorphism in our worries, as if trees responded to disease, age, death in the same ways as us. They don't. Disease is often no more than an irritation, to be neither cured nor submitted to but contained, isolated from the rest of the organism. Ageing and senescence are protracted stages in a tree's life in which it becomes more resilient and efficient. A beech can spend a couple of lively centuries in a so-called 'over-mature' state.

In similar ways we've transferred to wild trees notions about growth developed from agriculture and gardening. They'll suffer, we fret, if the 'fertility of the soil' declines. Trees certainly need soil as a source and reservoir of mineral nutrients, but not the deep and symbolically rich humus, that bath of nitrates and phosphates we often think of as '*the* soil'. This kind of enriched earth is uncommon in the wild, confined to river flood-plains and valley bottoms where plant debris accumulates. Trees evolved to make the most of poor soil, on which their slow-growing but durable and nutrient-storing trunks give them an advantage over lower, fast-growing plants. In rainforests, trees flourish on just a few inches of soil, and almost all the nutrients are circulated inside them, above the ground. In the Mediterranean, beeches grow modestly but to great ages on pure limestone. Most trees will grow better with more mineral nutrients than they're accustomed to in the wild, but find it hard to become established on the deep, nitrogen-rich humus of cultivated areas. Occasionally over-enrichment of the soil will kill them, by poisoning the network of underground fungi and bacteria with which most tree roots are symbiotically intertwined.

One friend of the beech, the 1930s rural writer H. J. Massingham, struggled with this paradox throughout his working life. He was an advocate of back-to-the-land living and co-founder of the Soil Association, and believed that earth (not the Earth) was the foundation of any proper relation between humans and nature. 'There is a kind of music in the order of the universe which penetrates man by and through the earth . . . In losing touch with the organic processes of the earth man is fouling the sources of his own being.' Massingham lived in the Chilterns for a while, and wrote impassioned words about their ancient pollards:

> To maintain the slow, ponderous, primordial life in themselves, they have agonised their woody, stunted, massive torsos into incredibly serpentine and elephantine shapes . . . they have pushed huge pachyderm roots which grasp the soil like a boa-constrictor wound about a stump, or like a mammoth planting down its foot. A suburban pleasure ground! – the proper companion for these trees would be towering saurians to rub their scaly flanks against them.

The 'suburban pleasure ground' was Burnham Beeches, the great wooded common only 20 miles west of London, where George Orwell's characters in *Keep the Aspidistra Flying* make their brief and famous escape from the smoke. Massingham's description of the connection between Burnham's beeches and the soil is unambiguous:

he sees them as claws, holdfasts, massive tendrils. This is a different kind of relationship between living organism and earth from that suggested in most of his writings. There is a telling photo used as the keynote illustration in his collection *England and the Farmer* (1941). It shows an anonymous land-worker's arm (he has no presence as a person beyond this) grasping a clod of the soil in which his feet are planted. The veins in his arm are heavily distended. He is rooted to the earth. Soil becomes blood, blood – the photo's title is 'English Earth' – becomes nation.

The beech is a good antidote to these notions, and to the unsettling moral and political ideas that sometimes accompany them. It haunts the kind of ground farmers have no interest in, the chalk hills too steep to plough, the acid marginal land, the gale-wracked mountains. Its life is more a matter of conjuring with light and water than dependence on the earth. If we have to have a cultural stereotype for it, it would be best cast as an air-plant, a tree of fancy and upward-lookingness, blown this way and that by circumstance.

Chapter Two: Heartwoods

The search for the ideal tree, the 'natural' beech. How has this coloured our view of what primeval woods were like?

I

IN 1787, JOHN WESLEY, charismatic preacher and founder of Methodism, tried to construct a three-dimensional sermon from beeches, a living model of how he thought the world should be. Not, I imagine, from any illusion that the beech was the Tree of Knowledge but because its suppleness makes it a tree of conciliation. He was on a mission to the industrial wastelands of Lisburn in Northern Ireland, and staying on the estate of his friends the Wolfendens, near Lambeg. It's an area beyond the natural range of the beech, but there were saplings growing in the garden, and Wesley took it into his head to twine two together, to represent the unity of Nonconformist and Anglican Churches that he longed for. For a passionate Christian, it was perilously close to sympathetic magic: splice the trees and marry the faiths.

But the beeches had agendas of their own. The pleaching took, and for a while there was an amicable intertwining, an ecumenical dance for two. But then dissent broke out. Breakaway branches shot off in contrary directions. Sects formed, turbulent eddies of argumentative wood, outlandish bossy gargoyles. Two centuries on, the Wesley Beeches aren't so much a simple parable, holding hands across the wildwood floor, as a rousing hymn to complexity and inventiveness. The point where the graft was originally made has now become a boiling mass of contorted wood and internal braces. The trunks above look disconnected from those below, two entirely new branchings. Born-again trees. Wesley's sermon has become a kind of creation myth, off on its own unruly business.

The entire history of our relationships with trees could be seen as this kind of debate. We argue them into forms that suit us, they respond with tortuous narratives of their own. We could win outright if we wished – and we do, repeatedly. But even amongst the most manipulative foresters there's always been a reluctance to see their successes as a kind of victory over trees. We need their growingness, their wilfulness, too much for that. They're symbols, witnesses, reminders of our own biological status as well as crude raw materials. So we resort to metaphors of improvement and stewardship rather than domination. We 'manage' trees and woodland to make them more successful. We put them out of their misery when they are 'geriatric' or diseased. During the 18th and 19th centuries both landscapers and landowners had the notion of an 'ideal' tree, a tall, uncluttered, classical column, grown from good stock and planted in fitting surroundings, pruned and pampered, and producing useful wood as well as satisfying contemporary conventions of beauty. The nurturing of the ideal tree perfectly expressed the idea that the outgrowths of nature were never entirely finished until they were subject to the improving touch of human art and industry. Some woodmen still use the word 'maiden' for an uncut tree: not yet ravished, but will probably be improved if it is.

One odd and obstinate consequence of the belief that human intervention is necessary for trees to fulfil their destinies is the modern myth that they grow only because they are deliberately planted. When they are cut down they must be replanted if they are to survive. Still the question I am most often asked if I show someone an ancient tree or wood is: 'When was it planted?' The idea that trees have successful reproductive system of their own seems to have passed out of the popular imagination.

In contrast to this has been the quest for the 'natural' tree, the 'virgin' forest, some arcadia free from the corrupting influence of humans. A place where nature could work out its own development plans. For much of human history this primeval forest, the wildwood, was seen in ambivalent light. It was the province of chaos and lawlessness as well as a symbol of freedom and pristineness. It was viewed with abomination and fantastical nostalgia in about equal measure. One potent myth was this: old Europe was cloaked with a boundless forest of immense close-packed trees, overwhelmingly oaks, which rose to the heavens and spread over all the dry land. Its trees never seemed to decay or die, nor ever had to put up with the ungainliness of youth. They were simply there, fully formed and immemorial.

Today, the idea of natural woodland seems both urgent and contentious. It's ethically desirable, scientifically fascinating, a grail for romantics and maybe the source of answers to a host of environmental problems. It's also entirely resistant to definition. Is the wildwood something lost in the past, or which might be re-created? Does it mean trees never interfered with by humans, or trees released from their interference? Is it another ideal, against which we can judge the merits of 'unnatural' woods and trees? And if 'naturalness' – however it's defined – is a desirable quality, what does this say about the role of humans in the ecosystem? The beech is a central player in this debate. No one is sure where it properly – 'naturally' – belongs. Its vulnerability to climate change seems an imperative reason for rescuing it unnaturally. And its plasticity of form means that it is hard to tell a rugged, manipulated, lopped beech from an ancient veteran of a storm.

Perhaps the most persuasive attempt to portray a natural beechwood is in Camille Corot's painting of *Forest of Fontainebleau* (1834). It's a forest caught at that moment in high summer when life does, just for a moment, seem to be in immemorial equilibrium. Fontainebleau is almost in the suburbs of Paris, but this scene could be the beginning of the world in a remote wilderness. The beeches unfurl upwards and away from a shaded glade in a bottom corner of the painting. They look benign, but Corot makes no concessions to classicism or the comfortable generalities of 18th-century landscape painting. These are real, believable trees. They're forked, split, tilting, ageing, rumbustious with branches. Their roots snake out through clefts in the rocks. They have the look of authentically natural trees, untouched by human hand. And between them, winding into the depths of the picture, is a low ravine with a stream in its bottom. It curls into the distance, a conduit for the eye between the immense sandstone boulders that are one of Fontainebleau's signatures – except that it goes nowhere. There is no conventional central focus, no church or great house or diligent ploughman, no crowning human presence. Just the glowing, horizonless haze of the countryside 'beyond'.

But at the very root of the painting, in the bottom left-hand corner from which the scene unwinds, a dark-haired woman, barefoot, open-bloused, is lying on her front amongst the flowers. She's wearing a russet gypsy skirt, and is described in the catalogues as a 'peasant', but she's lost in a book and looks casually sophisticated. By the 1830s Fontainebleau was already beginning to be a fashionable retreat for Parisian artists and bohemians, and Corot may have intended her to be one of those, an outsider, but utterly at home in this immense dappled enclosure: culture at ease with nature. It is a painting of perfect balance: the constantly changing patterns of light and shade, the human figure not dominating the scene but absorbed into it. A painting of the possibilities of humans as citizens of nature, not its oppressors.

Corot is playing with time-scales here, setting the modern woman in the ancient forest – the timeless wildwood. He was well aware that Fontainebleau had been lived and worked in for millennia, but chooses not to show it. And it is in that gap between the respective time-frames of humans and trees that myths flourish. If trees and woods had the same longevity as us, there would not be many tales to spin about about their origins and destinies. But the idea of a woodland that has existed in the same place for tens of thousands of years, constantly present but also constantly changing, is hard for us mortals to absorb.

And it isn't just that trees outlive us. Every stage of their lives is governed by rhythms alien to us. The vast majority of them expire in infancy. Those that grow into adults, the hardwoods at least, don't die when they're cut down, but sprout again – indefinitely, if they're allowed to. As they age they become progressively more resilient, and people plant them to try to prolong their own lives and influence by proxy. But between the planting, or appropriation, of a tree and its maturing, whole generations of humans and of cultural attitudes may pass. The forester's choice of tree is made irrelevant by changes in fashion or manufacture. The aristocrat's commemorative grove becomes a disorderly thicket as his lineage dies out. Whole civilisations may come and go in the lifetime of a single veteran hardwood.

The trees themselves are dense with time. They contain detailed records of their past, etched in the pattern of their branching and the texture of their wood. Wood is an accumulating memory bank, not quite like any other organic material. In the heart of a tree it's already partly dead. Its cells aren't constantly renewed, like bone or skin. But the trunk of a tree 400 years old contains tissue from each one of those years. Its experiences are engrained in it. Each annual ring of wood added to a trunk is a record of a tree's ecology of circumstance, and droughts, storms and insect attacks are registered in its width and texture.

Colin Tudge, in *The Secret Life of Trees,* imagines wood as an orchestration of the Four Elements, a conjuring of air, water, earth and fire (the sun's heat and light) into solid tissue. The Chinese call wood the Fifth Element because of this alchemy, but it does itself contain a fifth element, time. It's a matter of process as much as fixed substance. Wood, and the trees that contain it, changes with time as well as freezing it. This is the paradox we have such trouble dealing with, in our long envisaging of nature as a bulwark *against* change, as 'timeless'.

Images of trees change with time, too. Each age has had its ideal beech, regarded as an immutable archetype: the peerless source of fuelwood, the Palladian column, the wooden cathedral . . . The Florentine artist, Jacopo Ligozzi, the visionary botanical painter who once drew an angelica flowerhead from underneath, as if it were a parasol or a map of the stars, also made the first exact representation of a beech tree. Working in a Franciscan retreat in Piedmont in 1607, he made a drawing which gathered the tree's entire ensemble of physical possibilities and symbolic meanings into a single image. *The*

Beech Tree of the Madonna at La Verna is an immense single-trunked tree, broader than any beech you could see in the wild. The arched hollow at the bottom has the look of a retreat, or quite possibly Christ's tomb. The tree unfurls high above into the form of a rugged cross, a devout pollard, and in its cruck nestle the Virgin and Child, like roosting doves. Four centuries later, Hugh Johnson, doyen of tree connoisseurs, wrote of the straight and soaring beeches near Rouen (given their shape largely by French forestry techniques): 'One cannot help the feeling that the inspiration for the great Norman cathedrals came from the ancestors of this solemn forest.' What we believe to be the natural forms of trees, and the forms we press them into, reflect each other like a hall of receding mirrors.

II

When Sydney Parkinson, the young Scottish artist on Joseph Banks's ship the *Endeavour*, first set eyes on Australian woodland – doubtless the wildest forest he had ever seen – he could only make sense of it in terms of a prospect of the English shires: 'the country looked very pleasant and fertile; and the trees, quite free from underwood, appeared like plantations in a gentleman's park'.

My own first wildwood, my grove of initiation, was also a gentleman's park. Its squire was Graham Greene's uncle Charles, who'd owned the land on which our family house was built in the 1930s. The Greenes' seat was an austere Georgian mansion which dominated the eastern stretches of Berkhamsted's main street, and the novelist-to-be, already a fantasist, was a frequent visitor. He had a secret eyrie on the roof, where he'd gaze out loftily over the meadows and tree-land below: 'I would sit up there with my cousin Tooter, consuming sweets bought with our weekly pocket money . . . and discussing possible futures – as a midshipman in the Navy or an Antarctic explorer –

none of them to be realised, while we watched the oblivious figures in the yard and the stables from our godlike secrecy and security.'

Thirty years later the oblivious figures would have included our neighbourhood gang of ragamuffin incomers, all of us having much the same kind of dreams. The Hall and its grounds had been sold off for building during the great break-up of estates in the 1920s. But a sizeable chunk, including the landscaped area that lay closest to the big house, was left undeveloped until the late 1960s. It became our magic carpet, our gateway to the wilderness. We had no interest in how the place had arrived in its current state, but were relieved it was no longer controlled by adults, and was now *ours*. The Field, we called it, as if there were no other such place on the earth; it became our childhood common, a vision of a wild arcadia, a delicious tumbling-down of the pretensions of privilege. The old tennis-courts had turned into shaggy strip-lynchets. The remains of the Hall – hand-made bricks, marble sink-surrounds, scraps of prime oak beam, the grave-goods of the civilised life – lay in two huge long piles like Stone Age barrows. As for the parkland trees, those elegant embodiments of aspiration, they were in full-scale regression. They'd turned feral, broken ranks, developed subversively unkempt profiles. Forty years on I can remember them exactly as individuals, their look and disposition across the park. There was a sweet chestnut skyscraper, whose roots we camped in after it was toppled in a gale. An immense plane tree, a parasol for high summer. Birches we barked for kindling. Birches on a mound from which the first cuckoo always seemed to call. A stocky holm oak. A copper beech, but not many plain beeches. A walnut, whose nuts we hawked to homebound commuters.

But that personal map, annotated and indexed in my memory like a field-guide, is an adult's view, fixed in the last days before the park was built over. Back in the 1950s, I doubt I knew the names of half the trees let alone what they looked like as figures in a landscape. My own view of them was from the inside, as hiding-places, camp-sites, birds'-nesting opportunities, mines for firewood and spears. From that point of view beeches didn't get a look in. They were smooth-barked, unclimbable, of strictly scenic value. Trees for grown-ups. We made totems instead out of the great cedars of Lebanon which were scattered like the remnants of some post-glacial forest round the edges of the park. They were our compass points, the name of the place we lived in: Cedar Road! They had the aura of ancient monuments but none of their remoteness. We went up them like lemurs.

Two storeys up in a cedar was like being in the loft of an abandoned barn. The huge side-branches levelled out towards their tips, so that there were layers of springy floors, linked by a labyrinth of aerial passageways and covered with cobwebs and the choking dust of decades of trapped needles. We spent whole afternoons up them, posing on branches, gossiping, doing mindless things with pieces of wood, dreaming of the savage life. Then we'd go home for tea.

Then, one early morning in my late teens, long after I'd moved on from going native in the cedars, I woke to an unfamiliar whining coming from the direction of the park. I looked out of the window and one of the cedars had vanished, felled before anyone was about. A few years later – and another howling at first light – a second was toppled, on the edge of a local planning officer's front garden. Danger and inconvenience were given as the excuses, but in those days even ancient trees could be felled on a property-owner's whim. In a backhanded way these dawn raids – scheduled before most people were awake – were an acknowledgement of our ambivalent attitudes to trees, a way of doing the job with the least aggravation from the public. But decades on, the sound of chainsaws breaking through a dream still knots my stomach.

*

Does the development of our personal feelings about trees echo in some way the evolution of our cultural attitudes towards them? Animism and interdependence giving way to management and measuring? The hunter-gatherer maturing into the estate forester? Do we remake our images of them at different stages in our lives? And if so, when do we begin to have a historical sense, to ask questions about origins?

When I abandoned my backdoor wilderness in my teenage years, it meant not just a widening of my horizons, but a new perception of trees. They began to be shrouded in symbolic meaning. They graduated to the status of landmarks, lucky charms, almanacs. When I went on ritual plods round the home patch it was as obsessive a business as trying not to tread on the joins in paving stones. I was full of injunctions to myself. Hug close to the blackthorn hedge on the right-hand side of the lane. Gaze up at the ash where the first chiff-chaff sang. Cut off the corner by the sentinel beech. Scrabble under the turkey oak for a letter from my forbidden first girlfriend. I was beating my own bounds, sense-marking my territory, acknowledging – for reassurance that life went on, I suppose, even during that edgy time of adolescence – what I'd nodded at the day, or the year, before. William Hazlitt, confessing to his own rituals of walking, wrote of 'bending my eye forward, stopping and turning to look back, thinking to strike off into some less trodden path, yet hesitating to quit the one I am on, afraid to snap the brittle threads of memory'.

But beeches remained the remotest of trees for me. I gazed at them from a distance but rarely noticed them, or thought about what they were. One prospect that I went back to compulsively, as if I was searching for something I couldn't quite glimpse, was packed with beeches. But I saw nothing but a view of romantic England, or perhaps of my own over-cosseted spirit. I gazed at it in a dream-like state, a green study. The narrow valley wound away to the south, towards the Chiltern hill country, and some winters a thin stream rushed through it, falling out of the haze of trees on the horizon. Local legend had it as a woe-water, a sign of troubles past, or to come. But for me it was a thrilling omen, an intimation of the possibilities of life, and being in thrall to that idyllic vista was as close as I have ever come to a religious experience. What I didn't know then was that a smudge of green about half a mile down the valley, an indistinct swell above the cloak of hazel and hawthorn scrub, was a clump of ancient beeches called Heathen Grove. If I had known, I doubt that it would have stirred the slightest smidgeon of curiosity about the past of this place. Trees were part of what was present.

But I did absorb the proper adult view of the beech. Our teachers informed us that it was introduced by the Romans, who they regarded as the originators of all things cultured and interesting. Our parents talked of beechwoods as 'green cathedrals' – as if the Church were older than the Forest – and regarded the trees as the epitome of forest grace. They were elegant feminine foils to the rugged, more masculine oaks. Years later, I read a piece by the conservative journalist Paul Johnson (thinking perhaps of the Prime Minister whose cause he'd so ardently supported) lauding a pollard in the Quantocks – 'a queen-beech of quite astonishing complexity' – as a kind of arboreal Amazon:

> It was a brazen witch of a tree, a Lady Macbeth with more than a touch of both Goneril and Regan. I longed for a Disney to animate it and make it the anti-heroine of a gruesome Gothic drama, clutching and enveloping in its greedy branches some Bo-Peep or Snow White or Goldilocks and defying any benign green spirits of the woods to effect a rescue.

That was a good deal closer to my own feelings than the hushed reverence of my elders. But, to tell the truth, the beech was still a cipher, an abstract component of local woodiness. As the teenage romantic gave way to a rather earnest amateur naturalist, I spent more and more time amongst beech trees, but looking for what lived with them, not at the trees themselves. When I first discovered Gilbert White, pioneer of literary ecology and author of *The Natural History of Selborne,* I went down to his Hampshire village, and prowled around the hollow lanes. I walked in his beloved hanging beechwood – but with eyes down, peering for flowers. I found hellebores in exactly the same sites he had described two centuries before, and heard a nightingale in the distant scrub. Back home I haunted the beechwoods that were strung along the scarp near the Queen Beech, and became more proficient at identifying toadstools than the leaves of trees. I found dog's stinkhorn, with its absurdly lifelike pink prick; an earth-star, arched like a tumescent starfish over the leaf litter; fungi smelling of radish, aniseed, coconut (the so-called 'beechwood sickener'). And I saw my first beech-tufts, the porcelain fungus, the quintessential beech parasite. They grew out of the trunks like exudations of wax. They were pallid, slimy, ectoplasmic, almost translucent, and as they aged they began to droop like the clocks in Salvador Dali's eerie painting about time, *The Persistence of Memory.* Once I saw some sprouting from a beech which had tumbled into a river. They looked like glutinous water-lilies. But up in the woods my eyes didn't look any higher than the highest tuft, and the trees remained as backcloths. As I blithely wrote in an October diary entry: 'In woods you need to look closely at your feet as you do when mushrooming in meadows.'

I'd kept a nature journal since the late 1960s, a strictly non-confessional record of weather oddities, birds seen, plants in flower. There's no clue in it to what I felt about trees – except occasional hints of an excited stand-off between the old romantic and the new naturalist. When I first visited the New Forest, the most spectacular ancient beechwood in northern Europe, the fussy rationalist in me couldn't make head nor tail of it: 'Is it *countryside* at all,' I jotted, in severe style, 'or just a specialised habitat artificially preserved inside what is now the commuter belt?' I'm not sure now what kind of rigid categorisation of landscape I had in mind. But at least I was bewitched by the birds: buzzards wheeling over my hotel, lowland curlews bubbling amongst the heather, woodpeckers drumming on the beeches. It was as if the physical forest were a kind of stage-set, of significance only because of the action being played out inside it.

I know my feelings weren't quite as superficial, as oblivious, as that, but I'm not sure I trust my memory about them. Our attitudes towards nature are hedged about by expectation and second-guessing, and I could all too easily project the sentiments of his more reflective elder self into the head of that young bird-watcher. There's just one entry honest enough for me to recall exactly a particular January walk in the Chilterns. The words are minimal: 'Up in the beech-hanger at Ivinghoe. A keen breeze out of a distant ash-coloured sky. Haven't been out walking like this for weeks it seems, and the wind and space cut through my foggy head like an inhalant.' But I can remember the day precisely: the end of a rough Christmas, a spell of sickliness, that faint premonition of spring that comes sometimes on mild January days.

What I was feeling beyond that – not a sentiment fit for a nature journal – was thrilling: *no one in the world knows just where I am at this moment.* Beechwoods swallowed me up. They were like vertical mazes, countless subtle variations on the idea of 'tree'. They had a sense of spaciousness that was quite different from other woods. They weren't blocked by side-branches or clogged with ramping undergrowth. Often you could see into their depths. They were like caves. The silver-grey trunks were as smooth and glacial as stalagmites, and the light filtering through the half-transparent leaves moved over them like the sunlight on ripples. Sometimes I used to walk about in slow motion, imagining myself swimming between them. Beeches were becoming part of my embryonic idea of landscape.

That feeling of fluidity, of their being a kind of aquatic vegetation, has been a motif in writings about beeches. In the 1920s Sylvia Townsend Warner wrote of winter beechwoods 'dark and resonant as the inside of a sea-shell', and 'beech-hangers throbbing like sea caverns through which the wave had passed'. And the feeling of being deliciously submerged when I was in them grew. One of the Chiltern beechwoods' most expressive inhabitants was the tiny olive-coloured wood warbler, which spent the summer high up in the canopy. It's all but vanished from the region now, a victim of changing weather conditions over its migration route. But in my first published piece of nature writing, I tried hard to catch its presence.

> It was the first warm day of May and you could still see the sky through the latticework of branches. Primroses were in flower, and blackcaps and garden warblers singing in the undergrowth. And in their midst – the soloist to their chorus – was a single wood warbler, in full tremulous song. It was on a thin beech sapling not 12 feet from me, and even without binoculars I could see its throat shaking with every note of that falling, clear-water song, and seeming in the filtered sunlight to be the same translucent green as the young beech leaves.

My amateur attempt to make the wood warbler an almost amphibious embodiment of the beech was far from being the first. Gilbert White, who first distinguished them from their cousins the chiffchaff and willow warbler, listened to them in the beech-hanger in Selborne, where they 'make a sibilous shivering noise in the tops of the tall woods'. (The bird was later given the scientific name of *Phylloscopus sibilatrix.*) The poet Edward Thomas's beechwoods were only half a dozen miles to the south of Selborne. Early in the 20th century he wrote of the sharp, cascading song 'as if, overhead in the stainless air, little waves of pearls dropped and scattered and shivered on a shore of pearls'. To one anonymous writer it was 'an image of raindrops scattering among the leaves'. Lord Grey of Fallodon believed that 'the soft green and yellow colours of the bird are in tune, with the foliage, and its ways and movements . . . animate the beauty of young beech leaves'. For W. H. Hudson it was almost a beechwood dryad:

> It is a voice of the beechen woods in summer, of the far-up cloud of green, translucent leaves, with open spaces full of green, shifting sunlight and shadow. Though resonant and far-reaching it does not strike you as loud, but rather as the diffused sound of the wind in the foliage . . . a voice that has light and shade, rising and passing like the wind, changing as it flows, and quivering like a wind-fluttered leaf.

What none of us seemed to think of exploring was the meaning of the song from, as it were, the warbler's point of view. Far from being a creature solely of the 'tops of the tall woods', the wood warbler builds its nest amongst dead leaves on the ground, under the mantle of low sprays of foliage. Is its singing in the 'far-up cloud of green, translucent leaves' a quest for a kind of amphitheatre? Are its shivering notes designed to reverberate through the dense foliage? Tree and bird alike are pressed into human roles and metaphors, their own projects ignored.

For me the liquid quality of beechwoods reaches its climax at bluebell time. Bluebells, supremely shade-tolerant, are the signature flower of the southern beechwoods. No other kind of wood is flooded – it's the only word – with blue in quite the same way. In some places I know in the Chilterns the bluebells and beech leaves open in the same few days, and walking in the green-blue glow is like wandering through an aquarium. I read Gerard Manley Hopkins talking of shoals of bluebells 'wash wet like lakes'. In his journal, for 9 May 1871, he writes: 'in the clough/ through the light/ they came in falls of sky-colour washing the brows and slacks of the ground with vein-blue'. Two years later, on 11 May:

> Bluebells in Hodder wood, all hanging their heads one way. I caught as well as I could while my companions talked the Greek rightness of their beauty, the lovely/ what people call/ 'gracious' bidding one to another or all one way, the level or stage or shire of colour they make hanging in the air a foot above the grass, and a notable glare the eye may abstract and sever from the blue colour/ of light beating up from so many glassy heads, which like water is good to float their deeper instress in upon the mind . . .

'Washing the brows and slacks of the ground' – what is special about beechwood bluebells, so rarely diluted by any other kinds of flower, is the ebb and flow of them. They move like water – eddying, settling into flat pools, turning ice-grey at the end of spring. Even at their colour extremes there can be something aquatic about them. I once found a colony of variegated flowers whose bells, white for the most part, looked as if they had been dipped in blue water-colour paint, and then allowed to run.

I'm not sure if my sense of the beechwoods' watery aura was just an aesthetic conceit, or whether I was subconsciously beginning to glimpse something fundamental about how they worked – the slipperiness of life inside them, the glacial quality of their familiars as they unfurled themselves in the shadows and merged into the slow-flowing rhythms of the wood. There seemed to be nothing jagged about beech life. Sometimes I felt like a beech-creature myself, slipping through this deep ocean of sinuous shapes and muted colours, escaping more thoroughly from the world outside than in any other kind of place I knew. I was adrift in the woods, floated this way and that by currents in the trees that I didn't yet understand. Years later in a woodland sculpture park, I found a piece called 'the Fish Tree'. It was the iron skeleton of a tree with metal fish at the end of its branches, autumn-rusty against the spring green leaves. The odd thing was that it looked less like a shoal of fish than the real beech next to it.

And then in the mid-1980s I had a watery epiphany in the Beeches. It was autumn, and I was wandering about renewing old acquaintances, when I was caught in a downpour and scrambled under the nearest beech for cover. The rain was ferocious, spattering off the leaves in jets. The whole wood was changing colour, the trunks slicked into slate greys, next year's buds glistening like glazed fruit. And just a few steps away was the Praying Beech itself, half as tall as Nelson's Column, but blown to the ground four years earlier. And it was dissolving. It was lying flat out in the drumming rain and turning into soup, a dark ooze of charcoal and sponge-wood and fungus flecks. Huddled in my woody retreat, peering through the curtains of rain-wash, I felt I was gazing out at the beginning and end of things.

I'd been caught out by the deluge but not surprised by it. Since the mid-1970s the weather had been turning increasingly strange. There had been processions of droughts, heat-waves, wrecking winds. The trees in Frithsden Beeches were slowly going down, one by one, and the woodland had begun to look like a wooden henge, dotted with bleached memorials, their upturned root-slabs tilting into the soil. The supplication in the Praying Beech's clasped branch stubs hadn't worked. It had come down in spectacular style, half yanked clear out of the ground, half split across its one and a half metre diameter trunk. It was already rotten inside, honeycombed by fungus and charred near the heart, as if it had survived, and grown round, a lightning strike, or a bonfire started near the trunk. Bees had nested in the stump that summer – an old symbol of regeneration: 'out of the strong came forth sweetness'. And now, the whole of the rest of the trunk was embroidered by fungi. I couldn't put a name to the buff bonnets lurking in rot-holes and ranged in troops along the dead branches, but there were vivid splashes of colour – molten, chromatic colour. Purple beads of *Ascoryne*, pink coral spots, gelatinous folds of yellow brain fungus, white-tipped black claws of dead-man's fingers. It was like a reflection of the tinting of the autumnal leaves above, that prismatic separation of the trees' elements before the winter shut-down.

Then the trunk started to liquefy, in front of my eyes. The rain swept in from the south-west, hammering drills of water at the trunk. Flakes of bark began to fall off, then sooty dregs from the barbecued heart, and essence of beech dripped onto the woodland floor like the oil from an alchemist's still. The Praying Beech had turned into a Lazarus Tree, a parable of revival. Doused with water while it was lying flat on its back, it had sprung into a different kind of life. It felt odd picturing it as 'dead', watching it going through the rites of disembodiment in such florid style.

And it must have been at about this time that I first truly saw the Queen Beech, looked at it for itself, not just as a backcloth to the comings and goings of toadstools and birds. And I am fairly sure that, before this moment, I assumed that this hacked about mountain of wood was in an entirely natural state, and that this was how all trees appeared in their old age.

III

I'd like to say that this sense of beechwoods as labyrinths risen from the deep began to stir some curiosity in me about their origins, put me on the trail of the aboriginal beech. But again my journal holds a more pedestrian story. It was the autumn of 1973, and I'd been at a big conference at the University of Sussex on 'The British Oak'. It was a dazzling presentation by pioneers of what was the comparatively new science of historical ecology, and I went home with my head buzzing with images of woods. They shifted about in my mind like the figures in a kaleidoscope, changing form, glittering momentarily in the light, endlessly regrouping over time. And we were the ones who could shake them up.

Two days later I was up in the Ivinghoe beech-hanger again, with all the zeal of a convert:

> Have found a new eye for the subtle effects of light, moisture, soil, incline. Found a carline thistle [a chalk downland species] growing extraordinarily, *inside* the wood. A beech at the edge of the hanger had died, allowing in the best of the afternoon sun. When the sun broke through I could see the newly-lit area in dappled outline. There were other chalk grassland species – ploughman's spikenard, wild basil. Saw the whole hanger for the first time in terms of light: the wood at the top very old and dense, little light and no ground cover. Near the edge, where the light breaks in, that marvellous ground layer of woodruff, lily-of-the-valley, green hellebore, violets, spurge laurel. The edge scrub has whitebeam, wayfaring tree, young yew, and the grass

> bank just beyond, thyme, marjoram, felwort. Found the fungus *Xylaria polymorpha* [dead-man's fingers] on a stump, next to a discarded sock.

The schoolboy up the cedar tree had not entirely disappeared.

The star turn at the Oak conference had been a young Cambridge ecologist called Oliver Rackham, whose paper was a summary demolition of most of the myths about trees which had prospered in the age of commercial forestry. He described the continuous cropping systems which had been practised since the Neolithic period, and the capacity of trees to regenerate themselves, by regrowth or by seed. 'Contrary to popular belief, the harvesting of woodland produce did not destroy the wood. When a man felled a tree, he expected it to be replaced, normally without artificial planting.' He talked about the continuity of ancient woodland, and how these techniques – essentially the harvesting of 'found' wood – could maintain the structure of woodlands and their mixture of species in a way that replanting rarely did. Years later, in his classic book *The History of the Countryside,* he used the metaphor of ancient wood as text, and likened the treatment of the historic landscape to the ruin of an immense library of books:

> Many were written in remote antiquity in languages that have only lately been deciphered . . . Every year a thousand volumes are taken at random and sold for the value of the parchment . . . A thousand more are restored by amateur bookbinders who discard the ancient bindings, trim off the margins, and throw away leaves that they consider damaged or indecent . . . The library trustees, reproached with neglecting their heritage, reply that Conservation doesn't mean Preservation, that they wrote the books in the first place, and that none of them are older than the

eighteenth century; concluding with a plea for more funds to buy two thousand novels next year.

This all went to my head, and I turned into a kind of wood-crawler, out on my own hunt for intimations of antiquity, searching for the coded signs of continuity. Woods named after their parishes, not after some self-aggrandising owner. Woods with curious shapes and wavering boundaries, not the straight edges of planned plantations. Medieval banks and ditches. Old coppice-stools, as wide as small ponds. Suites of plants with poor colonising powers – wood sorrel, wood anemone, woodruff – that kept to the old woods where they'd first grown. I loved becoming aware, for the first time in my life, of natural structure, of the arrangement of older trees and young saplings, the changing patterns of light and density and species as you passed from slope to plateau, from dry soil to wet. All these characters and qualities gave each ancient wood a unique identity. They felt, I wrote at the time, 'like life-rafts out of the past'.

I spent as much time browsing maps as I did prowling through the undergrowth, dowsing the outlines and positions of woods for hints. I tried to spot patterns in scatters of woods that might once have been joined, scrutinised their names for hints of provenance. And that is how, poring over a large-scale Ordnance Survey sheet of my home range, I discovered that the green haze at the end of my teenage valley, the remote wood that I gazed at but never visited, had a name. It was called Heathen Grove. I couldn't imagine what this meant. A site of diabolical goings-on (there were many 'Dell's Woods' – i.e. Devil's Woods – round about)? A memorial to a plucky Enlightenment atheist? I badly wanted to see it, but getting in was another matter. It lay on the top of a steep ridge, barricaded on three sides by the barbed-wire fences of a notoriously irascible farmer. On the fourth was a long and almost equally impenetrable thicket of larch trees and thorny chalk-scrub. But that seemed the safer bet, or at least the more romantic. Luckily there was a track leading up the edges of the brush. I broke off the path by a badger sett, squeezed through hazel coppice, dropped to my knees to follow animal tracks through the thorns – and began to find what for me were prodigious plants. Twayblades and spotted orchids. Fly orchids, like wingless, velvet-bodied insects, flashed with metallic blue. A single clump of herb paris, with its crown of eight golden stamens. I discovered later that paris had been found on this estate 150 years earlier, by the owner, Augustus Smith (later to become the saviour of the Beeches on Berkhamsted Common) – a revelation of continuities that gave me gooseflesh.

A few minutes scrabbling, and the scrub began to open out. There were occasional young beech and ash and yet more orchids – white helleborines, though not yet with their elgant white flower-cups. Then I was in the Grove itself, surrounded by beeches and swimming through waves of fading bluebells. But its name still seemed inexplicable. By no stretch of the imagination was it a pagan version of a 'green cathedral', some dark and hushed retreat of soaring columns in which you might be tempted to tilt a reverent eye upwards in search of a dryad or two. It had cherries and hollies and large numbers of beeches, but they were striplings, a hundred years or so old, and light flooded into them from the open farmland to the west. And my eyes, as usual, were tilted uncompromisingly downwards, because what became rapidly clear was that the Grove possessed an astonishing ground flora, a collection of fugitive, old forest species which belied its apparent youthfulness. Amongst the bluebells were early purple orchids, unknown elsewhere in the parish. There were sheets of a flower I'd never seen before in my life – the coralroot, the lilac-flowered lady's-smock of the woods, which reproduces itself by means of curious bulbils, tiny beads of black growing in the axils of the leaves. And on the very edge of the Grove I found a few true oxlips, a hundred miles away from their Suffolk heartland. I'd discovered small colonies of these in some neighbouring woods the year before, and wondered now if those woods and Heathen Grove had once been joined. Breathing in the oxlips' slight apricot scent, I imagined I was making contact with the ancestral wildwood, with the relics of a population that might once have stretched through unbroken woodland all the way from the Chilterns to East Anglia.

As I walked back home, I stopped at the gate where I used to gaze out at this tumbling muddle of wood as a teenager. In my mind I stripped away the layers of growth I'd struggled through. The larches, planted by Italian prisoners during the Second World War. The dense scrub that had grown up on chalk downland, maybe in the agricultural depression of the 1930s. And what was left was just that lonely beechwood, sitting on top of the ridge at the head of the valley. The name now made some kind of sense. It looked like the kind of hilltop plantation popular with 18th-century landowners, hoping to echo the groves they fancied were once used for pagan worship. But if that were its origin, how did those ancient woodland plants get there? The transactions of the beech with time were still mysterious to me.

Yet, ironically, this revelation in my own heartland made me disenchanted with many of the other beechwoods I was beginning to explore. Few of them had the dazzling floral show of Heathen Grove. Some had precious few ground plants at all. They were monotonous, even-aged, unrelieved by other trees. I knew from maps that most of them were on the sites of ancient woods, but they'd been simplified sometime during the past two centuries, overplanted, replanted, thinned to favour trees suitable for commercial timber. I was sure that the trees in 'natural' beechwoods (whatever that meant) must have been more varied, a muddle of different species and ages and forms, like the grove round the Queen Beech. The timber beeches looked vulnerable, too. Their uniformity seemed to be magnifying their traditional misfortunes. They were packed close together so that their shallow root-systems were cramped, and were knocked over easily by gales. Grey squirrels regarded their dense and regular stems as convenience eateries. The growing numbers of deer were grazing away natural seedlings. On the fast-draining chalk soils especially, the beeches' meagre roots were having trouble coping with drought. By the mid-1970s, planners and foresters were beginning to predict a beech apocalypse, a mass collapse from these combined assaults – though, as a non-forester, I couldn't see what the problem was. These bland plantations were simply being edged towards what I regarded as the condition of naturalness.

From the late 1970s, the Chilterns Standing Conference's regular statements on the condition of the beechwoods became increasingly gloomy. It feared that bark-stripping by grey squirrels was stunting mature trees and killing younger ones. The repeated spells of drought were parching the older beeches, making them even more vulnerable to storms. The woods, it concluded, were 'ageing and degenerating', and there was 'a widespread belief amongst professional foresters . . . that over the next 10–20 years very extensive areas of unmanaged wood will be affected as trees die'. Some landowners were already moving out of beech-growing altogether. The answer was even more intensive management – replanting, scrub-spraying, squirrel control, fencing off natural regeneration from deer. Even with such help, 'it is likely to take 25 or more years before the trees that become established again begin to read as woods in the landscape; and it will take much longer for them to attain the majestic structure of the giants they have replaced'. It would be unfair to suggest that this was a yearning for the instant appearance of woodiness desired by many aspiring 19th-century landscapers. It was a worry understandably based on foresters' needs for 'good' trees, and the public's seemingly fixed notions of what a wood should look like. But as a judgement for the Chiltern beechwoods as a whole it seemed to me to indicate a certain impatience, an unwillingness to go along with the trees' own agendas.

One local woodland management group put it more drastically: 'The Chiltern trees are dying! The work must go on or the Chiltern woodland will rot away.' Just what might succeed the rotten trees was not clearly spelt out. Probably dense scrub, of intolerably long occupation. This might 'contribute something to the wooded character of the landscape . . . but at close quarters it will be no substitute for true woodland'. 'True woodland': what is this? Something like the old image of the primeval forest? A wood in a state of perpetual middle age, not permitted signs of either youth or dotage? Do the ancient dwarfed beeches in the Mediterranean mountains count? Are there prescribed minimums for height and girth? 'True woodland' sounded like a return of that eternally sought grail, 'natural' or 'ideal' woodland. It certainly wasn't, in the foresters' view, a wild and rampant growth of thorns, even if this did shelter young trees. The landscape architect Nan Fairbrother was dismissive of this upstart vegetation in her influential 1970 book *New Lives, New Landscapes*. 'Incipient scrub always lurks,' she wrote, 'only temporarily suppressed: it is the state of original sin in our landscape.' In this dystopian vision, the opposite of the romantic ideal of the immemorial wildwood, woodland mugged by alien squirrels and bashed about by un-British extremes of weather could only survive with continuous human vigilance. Scrub was not merely woodland in the making, it was the threatening new climax vegetation, the bleak future of unmanaged England.

IV

In the early 1980s I longed to have a wood of my own. Maybe this was a nostalgic yearning for a place to go feral in, like the abandoned park of my boyhood. But it was also a response to what I saw as an increasingly fussy paternalism towards woods, a hostility towards their potential wildness and independence, a deep-seated desire to control them masquerading as an anxious concern about their future. I wanted to see how a wood worked at close quarters, to test my hunch that a wood could flourish, in its own way, with only the minimum of human intervention.

I've told some of the story of Hardings Wood and its development as a community project elsewhere.* What I didn't mention, I realise now, were the stories of its beeches. I still found them awkward trees, not fitting into the canon of ancient woodland principles. But Hardings had them, in numbers. The wood was slung like a hammock across a dry coombe high up in the parish of Wigginton, half a mile from the village in one direction, and about three miles from the Frithsden Beeches in the other, up on the crest of the next ridge. If I'd climbed to the top of the Queen I could probably have seen it across the valley, a green tump with one oddly abrupt edge. It was two woods really, an old and a new. The ancient part, by far the largest, had both species of native oak, hornbeam, a good deal of ash, cherry, holly and hazel, mostly grown up from stumps and seed since the last war. There were some magnificent wild beeches, maiden trees, about 120 years old. A few had grown like candelabras in the dense shade of younger trees. There was a gangling coppice-stool, though maybe a rare example of beech suckering, as the American species does. And a single copper beech, a natural sport which occurs occasionally in wild populations. (It's not a good name: the leaves are pale plum, not copper.) In spring sunsets it glowed amongst the billows of green, a premonition of autumn. Next to this old wood was a plantation of 90-year-old beeches. It occupied a third of Hardings' 7 hectares, but I barely glanced at it in those early days.

*See *Home Country*.

Before I'd made the deal to buy the wood, I wandered endlessly around, dreaming about what I could make of it. In the dark of the densely packed trees of the old wood, there were tantalising glimpses of the plants that might flourish if only there was more light. There were small colonies of classic woodland flowers: sweet woodruff, yellow archangel, wood sorrel. There were isolated clumps of rare beechwood familiars. A tuft of the feathery grass known as wood barley. A few August-flowering violet helleborines, scarcely visible in the leaf-shadow of high summer, where their tapering flower-spikes glowed like amethyst. And a single stem of wood-vetch, an astonishing discovery, as the plant had only been seen once before in the Chilterns, a century and a half before. It wasn't flowering in the shade, but I knew its scented, lilac-striped blooms from elsewhere, and saw it as a talisman for the future.

A good deal of this ancient section had been gutted for timber during the war, leaving just a scatter of older trees. The dense natural regrowth, chiefly ash, had been interplanted with hybrid poplars by the misguided owner, tempted by a Bryant & May subsidy scheme to encourage the growing of matchwood. The poplars had turned sickly on the dry and flinty soil and begun to collapse. Now much of the ancient wood was a thicket of youngish trees and rotting pulp, with not a seedling in sight. It was, in pessimists' eyes, a dark and unfriendly tangle. It would have sorted itself out in time, as the trees grew older and died and gaps opened in the canopy. But faced with the opportunity to speed things up, I turned as impatient as the Chiltern foresters. I wanted a 'true wood' too. I wanted woodland glades and sheets of spring flowers and nesting nightingales. I wanted to see the next generation of saplings reaching skywards before I was embarked on my own rites of succession.

But when I first started recruiting volunteers to help 'look after' Hardings, I scarcely had a clue about what exactly we should do, except that I wanted the wood to conform to my expectations of a 'natural' wood. And if that meant doing some unnatural things in the process, then so be it. It was a slow process gathering helpers. When I talked about the project to the local primary school, I was astonished to hear that none of the children had ever been there, even though it was only a quarter of a mile from the village – something that would have been unthinkable when I was their age. Most of the village adults hadn't been inside either, but still worried that we might be about to change things, cut trees down, bring in coachloads of visitors from the surrounding towns.

In the end, we had more outsiders than locals for our first working party, most of them friends who'd had experience of woodland work, even a few professional conservationists. Not really having any exact theories about what should happen next – except a conviction that we should do *something* – we resorted to a kind of on-site democracy. If we didn't yet understand the wood's own point of view, so to speak, at least we could have a good variety of human opinions. So we earnestly debated every action. We believed the wood need thinning, but where and how? Which trees to keep and which to take away? Some volunteers regarded bringing light to the pond as a high priority, so we began there. Others wanted a glade, so the pond edge grew into a clearing. We hacked sycamores and poplars down with abandon, but no one would touch the dark and sacrosanct holly, for fear of supernatural reprisals. The felled trees were logged and stacked in piles, to dry out.

What we were doing seemed enormous and presumptuous at the time. But from timidity and lack of skill, we did it with a rather ham-fisted delicacy, as if we were picking up a new-born for the first time. We were grooming the trees more than weeding them out. And next summer we went up to the tiny sunlit circles we'd cut, expecting them to be full of new seedlings. They were, but the minute ashes never grew taller than a finger, and in a couple of years the holes in the canopy had grown over again. I dreamed up a glib epigram to try to balance our enthusiasm with a touch of humble realism: 'We may be the accelerator, but the wood is the engine.'

With hindsight the modesty and chaos of our work seem rather apt. It was on the same scale as natural disturbances in the wood: local gales, a touch of honey fungus, the browsing of deer. Some of our first small pathways followed the lines of badger tracks. The rest seemed to occur spontaneously, following routes through gaps in the undergrowth or unconscious prompts from trees and patches of light. Most of the social and economic transactions in the wood were also small-scale. The logged-up firewood went chiefly as a perk to our helpers, provided they took it out by hand. We sold the excess to local pubs, and took a free load to the Peace Camp at Greenham Common. Small parcels of moss from the dead poplars were harvested by flower arrangers. Twiggery was destined for pea-sticks. I once worked out that 500 person-hours of work produced 25 tonnes of firewood, which at the current prices meant a return of 50p per hour, but that took no account of all these unquantifiable harvests.

And if the young trees were indifferent, at least the flowers responded to our beaverings, and to the new glimmer in these little clearings. Wood anemones spread throughout the wood, and on warm spring afternoons proved that, *en masse*, they had a scent, like warm honey. Spurge laurel, our native daphne, first flowerer of the year, broke out of its redoubts under the old beeches. The wood-vetch stretched itself along the paths we'd made, revelling in the new light and disturbed soil, and then broke into glorious sweet-pea-scented flower. At the end of three years' work, we'd clocked up twenty-eight of the species which were the specialities of ancient woodland in the Chilterns, a remarkable number, we were assured, for a wood of only 7 hectares.

But next to this naturally sprung heart of the wood was the beech plantation, about a hundred years old, but of unknown provenance. The trees were magnificent. They had never been thinned, and rose to immense heights. To foresters – and several came to inspect the wood – they had the look of promising timber. But my own view was more baleful. When I took visitors round, we went no more than a short way into this section, just far enough for me to make disdainful gestures at the impoverished life of plantations. When we began working in the old wood we ignored these soaring columns. They were out of scale with what we were doing. Too remote. Trees for grown-ups, as they had been when I was a child. I toyed with the idea of felling the lot, cashing in their timber value, and coaxing the old wood to reclaim its ancient territory. By my set of values the beeches weren't 'true woodland' at all.

Not everyone shared my sniffiness towards them. When the local primary school asked if they could hold their Ascension Day service in the wood, it was the beech plantation – the green cathedral – that they chose. Every May, the children tramped half a mile across the fields, gathered into a glade amongst the freshly opened leaves, and sang hymns to new life and the mysteries of transubstantiation. Then they were let off the leash for the day and ran riot through the trees, little angels turned into fox-cubs.

But the plantation remained a flowerless place, save for a scatter of bluebells near the paths. And, mindful of the fact that we might want to take our evangelical mission there one day, we created a proper forest track to it, up through the old wood. We felled a few poplars to mark the line, then had the village children in to dig up primroses and ferns for fostering during the work. All kinds of people added their special skills and perspectives. Len, the digger-driver who came in to excavate the track, used to be a watchmaker in London's East End, and he gave our uninspired route a serpentine twist, like a spring, that meant the late evening sun shone right down it. In March the children brought their potted ferns back, and added them to the surge of regrowth that had already started along the edges of the track.

Working in the wood was always a heartening experience, though often humiliating. I learned quite smartly how little I knew about the way trees lived, and how they responded to human interference. I was stuffed with mythical beliefs and hubris. I assumed that shade-loving hornbeams would pollard well in the shade. They didn't. I imagined that an area of rich, damp soil in the very bottom of the wood, quite devoid of trees, might be the one place where some small-scale planting might be justified. The trees knew better, and promptly died, suffocated by nitrogen. I'd been sceptical of management, and ended up doing it myself.

I was acquiring skills I'd never dreamed of. I learned how to use the chainsaws I'd once detested, how poplars split when they were felled, the kind of places that wood goldilocks liked to grow (in half-shade, along the edge of the wider tracks). But I found that work, and maybe the work ethic, could narrow my view of nature as much as extending it. Sometimes, trying to enjoy an uncomplicated stroll around the wood, I found I was trapped in a kind of managerial tunnel vision. I'd peer obsessively up at the canopy to see how much light was coming through, scour the ground for tree seedlings, check our few wild beech saplings for squirrel damage, make lists of jobs to do. I needed to make a conscious effort to relax, and remember that the wood had rhythms and priorities of its own.

But it was worth it when I did. I began to discover favourite patches where I'd sit down and try to absorb what was happening, in that spot, at that moment. This was my favourite: a small patch of thin, acid soil under some wild beeches and sessile oaks, the tilt of a half-vanished wood-bank, a single dark holly. The still centre of the wood. In spring the anemones kept to their place, down amongst the leaf litter, and the swaying wood melick to its, up on the bank. There were a few scattered bluebells which never reached more than 10 centimetres in height, and wisps of foreshortened honeysuckle. The only dramatic change I saw here in ten years was when the leaves of one bluebell, normally content with spiking dead beech leaves and bearing them into the air like ruffs, broke through the rotting shell of a fallen branch. Up here this was a major event.

I knew this minute oasis of stability wouldn't last, but I liked to sit and imagine the wood turning more restlessly round it. In winter it could turn dramatically. Gales raked through the beech plantation, tearing the dead wood out of the tops of the trees and spearing it into the ground. The place would seem full of new light, pouring down through the canopy and reflected off the gashes on the broken branches. Days of deluge were followed by sudden freeze-ups, and old logs, sodden as sponges, split open as the water bulked up into ice. When snow fell it bent the low outer branches of the hollies to the ground and welded them there, forming a kind of skirt. Woodcock sometimes rested in these shelters, and would shoot out at the last minute, like a stifled gasp. They skimmed crankily over the snow-dusted brambles until their chequered plumage merged with the shade.

Autumn gales and winter storms transformed the interior of the wood. Drifts of leaves or snow obliterated paths and left enticing new trails of wind-bared ground between the trees. I walked about as if I'd never been in the place before, seeing the other sides of familiar trees. One afternoon, scuffing my way up the track we called the Steep, I glanced sideways and saw what looked like a straight avenue tacking at a diagonal up the slope. The leaves had been blown hard up against the north-facing trees and formed a kind of banked track. I couldn't believe it was accidental. There were several large beeches and the oldest hawthorn in the wood in a dead straight line along it. For a moment I was seized by the conviction that I'd discovered the route of an ancient pathway I'd been trying to find for years. It was marked on an early 19th-century map, and was the chief track from the village down to the valley coach-road, passing through the wood on the way. But in 1853, during the enforced parliamentary enclosure of the whole parish, it was stopped off, together with twelve other old roads and footpaths. The notices of dispossession were posted about the district in the same heavy type used by the police for 'Wanted' posters. The eradication seemed to have been total, and no trace of a path survived where it had been marked on the map. I suspect my 'discovery' was a fantasy: any long-established trackway would have left a mark on the ground as well as through the trees. But it gave me a brief vision of a more sociable wood than the inhospitable one I'd found when I first tramped through it with an estate agent's sale-sheet in my hand.

But I think we did add something to the place, beyond brief stabs at the accelerator pedal. We rescued some of its almost vanished plants from probable extinction, giving them enough light to flower and seed. We created some of the habitats that occur quite naturally in large woods – ponds, steep banks, glades – but which have often vanished from the unnaturally small islands most woods have been reduced to. And I think the experience added something to the lives of the people who worked and played there: moments of real intimacy with living things, and opinion-changing experiences, like the witnessing of young trees springing unprompted from the ground.

I did, over the years, learn more about the history of the place, and the people who'd worked in it before us. Harding was the name of a 16th-century inhabitant of Wigginton, perhaps the wood's owner, or a woodman-tenant. The earliest map I was able to trace, from 1766, showed our wood as part of a bigger tract of forest, a mile and a half long and a mile wide at its broadest, stretching east almost to the boundaries of Berkhamsted. By the beginning of the 19th century, two-thirds of this great wood had vanished, turned into more profitable arable and pasture land. Just five small island fragments remained, of which Hardings was one. The tithe commutation map of 1841 showed the 20th-century outlines of the wood quite clearly, except that what is now the beech plantation was a field. Twelve years later the entire parish was enclosed by parliamentary enactment. The wood survived, but the dense landscape of commons and hedgy fields around it was devastated. The following year, a new curate was appointed to the parish, charged with the job of saving the souls of the dispossessed natives of this barbarous woodland community. He was aware of its local nickname, 'Wicked Wigginton', and described it as 'a *terra incognita* in the neighbourhood'. But his report to the diocesan bishop on the consequences of enclosure – kept secret for 150 years for fear of offending local families – rather took the villagers' side. 'All the picturesque appearance of the place was gone', he wrote:

> and perhaps the poetry. Post and rail fences were right and left all over the place. But I believed that the change would morally tend to the benefit of the people: they would be less rough, wild and uncivilised. I cannot judge whether this has happened; whether as some predicted 'wicked Wigginton would become virtuous Wigginton' . . . But as I look back I am much surprised that they accepted the enclosure as patiently as they did, considering of how many rights the enclosure deprived them.

At this time Hardings was owned by a woodman called John Garrett, who leased out a thin strip of land adjoining the wood, about half a hectare in extent, to a local smallholder, Mr Burch. On maps it's called, sardonically, 'Hundred Acres'. This strip has now turned to woodland. But there's a wide-branched beech pollard at its edge, its expansive shape a record of the time it was growing at the edge of an open field. Then the wood was bought by the Rothschilds, who owned large tracts of land in the area. It was during their tenure that the beech plantation was started, with Scots pines as a nurse-crop, a few of which still remain. When they sold up in the 1920s, they left one other memento – a right, inscribed in the deeds, for villagers living in the estate cottages nearby to discharge their sewage into the wood. It was the outfall from these pipes that caused the sludgy patch at the bottom of the wood where every tree we tried to plant died.

After the heavy felling in the last war, the wood passed into the hands of an absentee wood merchant, and was pretty much abandoned, save for the bizarre poplar plantings. Then I bought it in 1981, and passed my judgemental eye over the whole lot.

If one's evolving personal attitudes towards trees recapitulate historical relationships, then I reckon I'd reached the mid-Victorian age. I'd been a make-believe (though highly practical) hunter-gatherer, a teenage animist, a young Romantic, an earnest Enlightenment naturalist. Now I was a rather bossy, propertied antiquarian, who fancied himself a woodland connoisseur. What this last stage represented, I can see now, was a quest for authenticity. I had joined the quest for the ideal, the primordial, for 'true woodland' – and I had my own image of this, just like the countryside planners.

I'm sure I still have my own mythical beliefs and irrational convictions. But twenty years of being close to one small wood has punctured a few, and especially the notions that woods are static entities, or entirely dependent for their forms on what humans do to them. Hardings repeatedly went off on its own project, regardless of what we did. Beeches refused to regenerate in the gaps we cut for them, and their seedlings rooted themselves instead in patches of dapple close to their parents. Scarce and supposedly immobile fern species ramped through the wood along trails made by our pick-up truck. Rowans died for no apparent reason. Hollies spread about, turning into suckering trees where frost had welded their branches to the ground. Ashlings sprung up everywhere, voting themselves in as the wood's next dominant species. I had a map of every minute detail of the wood in my mind, but it went out of date every time a bluebell pushed through a rotting log.

V

Is there a convincing plan, anywhere, of what a primeval beechwood looked like? Boubinsky Prales is a virgin forest in the Czech Republic's Sumava Hills, a towering stand of beech, spruce and silver fir. It is reputedly one of the few untouched remnants of the European wildwood. No trees have been planted and no dead wood taken away. In 1847 an ecologist called Joseph John made an exact map of the standing and fallen trees in a sample plot of a few acres. It's an extraordinary bird's-eye view, like a plan of a spillikins game – except that lots of the sticks are vertical. The fallen trees are drawn to scale. Most are very straight and long, though there are some shorter logs too. Some of the trees are shown with their roots out of the ground, rinsed clean of earth. A few are resting against standing trees, probably not yet entirely fallen. There are 43 fallen trees and rotting logs, 38 large standing trees, and 190 smaller trees – much the kind of proportions usually found in old forests.

In 1868, Boubinsky was made into a state reserve. And in 1954, John's plot – and all its standing and fallen trees – began to be regularly resurveyed. And a picture has emerged of the wood not as some kind of fixed memorial but as a process, with time-scales and rhythms of its own. The old trees decay laboriously: sometimes like slow-burning tapers, rotting where they stand, sometimes snagged up against neighbouring trees, until branches break and they crash to the floor. The young saplings are slow-moving, too, becoming established in small groups where a collapsed tree has opened the canopy, but still in sufficient numbers to renew the wood. And, as in so many similar forests, beech is slowly becoming the dominant tree.

Not all images of Boubinsky Prales are as unambiguous as John's map. 'Prales' is a general Czech and Slovak term for virgin forest, but the locals have a rather different understanding of 'virgin' from professional ecologists. They dissent from the notion that it means a total absence of human activity. Of course there had been a bit of hunting and cattle-grazing. Of course peasants from the surrounding villages took out bundles of firewood. They had to survive in winter too. How could humans behaving like forest animals compromise the pristineness of their forests? Since the fall of the Iron Curtain, Boubinsky Prales has come onto the tourist circuit. A trail, marked with a low-railed fence, zig-zags through the wood. Its route is determined by the positions of the fallen logs, and is a maze of dog-legs and detours, taking you through a wood which, to an English imagination, is more densely packed with thin trees than you expect. At least the fences help protect the reserve's special assets. But it is a disorientating experience, keeping to the dotted line through one of the wildest woods in Europe.

This is the paradox of the mapping of nature, and of the contrast between human time and tree time. We are, by our own natures, namers and systematisers, compulsive searchers for pattern. We have evolved as this kind of creature. We can still follow our noses in woods, listen for noises in the undergrowth, keep the sun behind us. But our culture won't go away. It's what prompts our curiosity about how woods work and where they came from. It frames the kind of answers we find, and the stories we tell about them. Yet the minute that trees are imaged – defined, charted, conserved, logged, sampled, trail-marked – they are, to one degree or another, frozen in time. They are seen with the momentary exactness of a flash photograph. At worst they can become museum exhibits, pinned down like butterflies on a board. At this point even the idea of 'naturalness' itself becomes a fixed agenda, a checklist of qualities.

What is known about the early history of the beech in Europe comes from these kinds of snapshot. Trees leave traces of where they have lived, physical evidence that is datable by its depth or surroundings – fossilised leaves, identifiable fragments of charcoal, preserved dustings of pollen. Pollen is long-lasting and hard to destroy. The germ of the plant's future outlives all its more disposable parts. If it's blown into a peat bog or the silty mud of a lake it can be preserved for tens of thousands of years. Beech pollen is sparse and heavy compared to other trees', and doesn't often drift far from its origins. If there are beeches near a lake, there may still be very little pollen in the layers of mud. Beeches have, consequently, a fugitive presence through prehistory, not always leaving even a momentary record of their existence.

But at the end of the last glaciation, between 10,000 and 7000 BC, beech left traces of its pollen right across southern Europe. In the cooler temperatures, it grew at lower altitudes than it does today: at 300 metres above sea level in Provence, at 200 metres at the Lago di Martignano in Italy, at 30 metres in Saleccia, Corsica, at sea-level in parts of Greece and throughout the north-east Adriatic region. But there's little clue from the pollen as to how it grew, squat and spreading or tall and straight, in closed woods or more open savannah, mixed with areas of open grassland and scrub. Shrubs like cistus and lavender leave minimal pollen records, and the slightest trace may indicate large areas of scrubland amongst the trees.

As the climate warmed, beech began to spread north and west from these refuges. It reached the Languedoc in south-west France, and survived there until the current hot, dry Mediterranean climate set in about 2000 BC. In much of central and eastern Europe it became, for a while, the dominant tree. By 7000 BC it was on the sites of what were to become the Forêt de Fontainebleau, south of Paris, and the Forêt de Cerisy further to the north-west. From there it was just 120 miles to the future south coast of England, then still joined to the continent. The term 'land-bridge' isn't very helpful here. It gives the sense of a thin causeway between Dover and Calais, neatly snipped by the nascent English Channel. The separation was on a broader front than this, and more protracted. A wide strip of low-lying forest and saltmarsh, stalked by hunter-gatherers, would have joined modern-day Hampshire and Dorset to the Normandy coastline. The beech was the last of the great forest trees to make the crossing, in the wake of oak, alder, lime, elm and ash. It made it in the nick of time, about five hundred years before the Channel opened in 5500 BC, thus qualifying as a bona fide English native. It was remarkable progress for a shade-loving tree whose seeds aren't blown by the wind.

Its arrival date was also at least a thousand years in advance of the first wave of farmers from the Mediterranean. The earliest pollen deposits occur in the gravels and sands around what would become the New Forest, from whence it spread north and west into Dorset and Somerset, parts of Devon and South Wales, and probably the Sussex and Kent Weald. There seem to have been small outlying beech domains, too. Prehistoric traces have been found in East Anglia's sandy Breckland, parts of West Yorkshire, the Lincolnshire Wolds. Oliver Rackham has even seen ancient coppice-stools near Durham, big enough to date from before the beech-planting era.

Given the relatively benign climate, it's not clear why beech didn't spread even further, into areas where, as a planted tree, it's perfectly comfortable today, such as the Midland Plain and the bulk of the West Country. Its temporary halt is one of the tree's mysteries. The wildwood into which it tried to insinuate itself wasn't homogeneous, or static. It had become established in a landscape full of grazing animals and natural wetlands, and should have had sufficient breaks to allow beech further in. The tree seems to have no difficulty in penetrating an already established forest cover today, and it might be valuable to discover what lay behind that prehistoric diffidence. Maybe it was a hesitation in the face of early climate change.

But where the tree did become established it helped form the only communities which can unarguably be called natural beechwoods, existing before organised human intervention began. Pollen remains suggest that these were more mixed than the woods we're familiar with today, with lime, oak, ash and elm amongst the beech's companions. But what these original beechwoods looked like, and how they behaved, is elusive. Did the different tree species grow in exclusive patches, or intermingled with one another? Were the trees closely packed, the dense canopy only partially broken when several tumbled down together? Or were the openings larger, more permanent, kept open by browsing animals? How long were the periods during which beech ruled the roost? What was the individual tree's longevity? Were there early beech diseases? Did the self-replenishing forms seemingly invented by humans – the bushy coppice and the mop-headed pollard – already exist in embryonic form, as natural archetypes developed in response to weather and browsing? Whatever its state, the beech wildwood in England existed only for a thousand years. In 4500 BC the first settlers from the Mediterranean arrived, and began cutting down the trees.

Then, about 500 BC, beech rapidly expanded its range. It began invading chalk and limestone soils, in the Cotswolds, Chilterns and southern Weald. The pattern of the tree's spread seemed to follow the clearance of these light soils for agriculture, then their abandonment during a period of wetter weather that set in at this time. But this raises more questions about how beechwoods 'became'. Beech today tends to be a fifth columnist, expanding through existing woods from the inside, rather than invading from beyond.

Yet however it arrived there, beech was well established in the Chilterns at the beginning of the first millennium BC. The lineage from which the Queen Beech is descended began at this time. By the start of the medieval period recognisable Old English names for the tree appear: *boc, bece, boke,* echoing *buche* in German, and *beuk* and *book* in Dutch. Etymologists have customarily linked these words with the very similar roots of 'book', meaning a collection of writings, because the first European texts were inscribed, like literate graffiti, on beech trunks or tablets of beech bark. (A thousand years later Gutenberg was to cut the first printing plates on beech-wood.) But the Oxford dictionarists are sceptical, and it's sensible to be cautious about seductive etymologies. John Collier's surreal *Alphabet for Grown-up Grammarians* ('By a Supposed Lunatic', 1778) gives a suggestion about the origins of the word wood itself: 'For instance, take up a stone or an axe and knock it against a tree and I cannot help fancying but with the breath of the stroke it says *Wood!* as plain as the letters can form it or we can pronounce it . . . this I call the language of *God,* of *nature,* of *common sense.*'

Perhaps it is too common a sense that popularly interprets Buckinghamshire as the county of the 'buchs' or beeches (historians prefer 'the land of Bucca's people'). But the woods of the Buckinghamshire Chilterns were about to be invaded by natural language in any case, as large numbers of axes were knocked against large numbers of beeches.

Chapter Three: Wooden Walls

The 'most lovely of forest trees' was, for at least three thousand years, a tree to stoke fires with. It was medieval London's North Sea gas. It was useless for building ships and was left alone during the 17th and 18th centuries' frenzied naval arms race. But the need for fast-wood outlets led to the invention of the plantation, which was to change the landscape for ever.

I

IF YOU TRAVEL east towards the Chilterns from the Oxford plain, you see the scarp from five miles away, a dark ridge of beech-clad chalk stretching from one end of the horizon to the other. A green wall. From this viewpoint you can easily believe it is the second most densely wooded area of Britain. These days I mostly arrive from the opposite direction, so this prospect is a rare treat. But it's also, I know, a distorting one. For centuries the Chilterns was seen as a possible – and possibly dangerous – remnant of the wild forest, a south country Sherwood. The Chronicles of St Albans claimed, probably wishfully, that just before the Norman Conquest, 'to provide safer roads [their abbot, Leoftsan] cut back the dark woods which extend from the margin of Chiltern [*a limbo Cilitriae*] almost to London . . . for at that time there abounded through the whole of Chiltern extensive, thick and abundant woods'. In 1627 the traveller John Speed wrote that the Chilterns 'were so pestered with Beech that they were altogether unpassable, and became a receptacle and refuge for theeves, who daily endamaged the way-faring man'. William Gilpin thought they were the haunt of 'banditti'. All views from a cultural distance of rather more than five miles.

More intimate glimpses give a different perspective, a sense of the beech as protean, hugely adaptable. If I'm coming back from the east I pass through the woods round the Queen Beech: huddles of squat pollards by the side of the road, drifts of younger beech, birch and thorn sloping up to the common. From the west, looking towards Hardings Wood, I drive past crest after crest of storm-combed beeches, ragged staves of skyline trees with soft counterpoints of ash rising beneath them. Best of all is a favourite lane in the deep south, which climbs up towards red kite country through a beechwood in a ravine. It's a commercially managed wood, but the deep-cut slopes transform the trees, immersing them in layers of flickering leaf-shoals. In spring it's a cascade of beeches, the pale trunks breaching through the green surf.

Another wood, different again. Down in the deep coombes of Maidensgrove, north of Henley, there is a wood of beech bushes. They look like dark hazels, short, rotund, multi-stemmed trees bristling with new twigs around the smooth stems. These are epicormic shoots, probably the result of constant deer-browsing. But the larger stems may be a response to cutting. The bushes look as if they are what is called coppice, last cut in the Victorian period perhaps – though there seems to be no local memory of such a tradition. Beech doesn't regrow well after cutting in modern England, perhaps because of the depredations of deer, perhaps because late frosts damage the shoots. Or perhaps because the practice was discontinued. Done over many centuries, on lineages of naturally regenerating trees, regular cutting would have selected those trees with the strongest, most prolific regrowth. Trees with this form – and many grown much bigger – survive all over southern Britain, from Devon to north Norfolk. The Maidensgrove beech bushes have a striking local accent. There is almost always one thicker central stem in the middle of the younger regrowth. Did woodmen leave this as a lifeline for the tree, in much the same way as they did with hazel in Sussex? Some established growth to see the tree through the chills of early spring?

The regular harvesting of growing wood began at much the same time as the more permanent clearance of wood to make room for agriculture. Any hardwood tree that is simply cut to the ground rather than being uprooted will sprout again from the stump. So will trees that have been cut back by frost or broken by gales or eaten by grazing animals. There are dormant twig buds all the way up young trunks and shoots, which spring into active growth when the live wood above them vanishes. Hardwood trees have evolved with the capacity to regrow after they are damaged, and can seemingly do this on an almost indefinite basis. The form of tree that results is ubiquitous in both wild and managed landscapes: the roughly circular base, knobbly with scar tissue, the multiple shoots, sometimes seeming like independent trunks if the last cut happened a while ago, or emerging sideways from the stump (or 'stool') then curving back towards the vertical. Animals may gnaw away this regrowth, but it takes a lot to kill a well-rooted tree.

When this cutting-back is done deliberately, on some kind of regular cycle, it's known as coppicing. Something between five and fifteen years is the traditional interval between cuts. Shorter rotations give thinner brushwood, useful for fuel but not much else. Longer intervals can produce poles broad enough to be used, for instance, in rough and ready furniture. The final end-product is less significant than the process: the coppice is the original renewable resource, producing rough wood for whatever you need.

Neolithic peoples were the first to exploit this natural regeneration in an organised way, for fuel, cattle-fodder, building materials. The remains of ancient wooden trackways in swampy areas suggest that wood was being deliberately grown for specific jobs. The Sweet Track in Somerset, one of the oldest and most sophisticated, contains poles that are too straight and uniform to have been selected from casually cut scrub. The logs of oak, ash, alder and holly seem to have been chosen to fulfil different functions in the structure. They are remarkably similar to poles cut on a long rotation in modern coppice woods. The Sweet Track has been placed by radiocarbon dating to somewhere between 4000 and 3000 BC – a period which coincides with the settling in of the first farmers.

There's no reason to believe that beech wasn't coppiced here along with other trees – though the Neolithics would have soon discovered that, good as it was for fuel and fodder, it was useless for outdoor trackways. And, because it decays so easily, it was almost never used as a house-building wood. But it served as an all-purpose domestic wood, burned as fuel and hewed into bowls and rough furniture. By the medieval period its value as a producer of fine charcoal was well known, and it was in industrial use across Europe.

The fully developed coppice, where the regrowth was cut on a regular rotation of about ten years, and numbers of timber trees – 'standards' – were permitted to grow amongst them till they were 80 to 100 years old, was the first kind of managed woodland to have no equivalent or prototype in nature. Some aspects of it echo natural processes – the regrowth from the cut stumps is similar to that from trees munched back by animals, the cut areas to natural glades – but they're accelerated, compressed, regularised. The flush of light every ten years does wonders for spring flowers and regenerating trees, but at the expense of two-thirds of a wood's natural life. No growth above the ground is older than 100 years, or taller than about 15 metres. Plants and animals that depend on high canopies, on old and decaying wood, or on permanent shade, simply don't survive in worked coppice. These are all important components of 'natural' woodland. Permanent shade may seem inimical to plant life, but there are whole groups of mosses and fungi which are dependent on it. Dead and rotting wood provides crucial habitats for insects, under bark, in rot-holes, ranging along the dried-out tunnels inside hollow trunks. And since 'fallow' wood may make up half the woody mass inside an unmanaged forest, large numbers of organisms have evolved as dependent on it. There is usually next to no dead wood in a worked coppice, nor any oasis of shade which lasts more than a decade.

Coppices are an ingenious human invention, inspired by a natural process. They sustain some of the tree species of the wildwoods from which they derived. They provide a habitat for creatures of the light, and a sustainable source of wood for humans. But they are only aspects of woodland, and far from 'natural'.

Beech coppice flourished across Europe, as one of the best sources of a highly calorific fuel. It was cut in the humid valleys of Sussex, and in the high mountains of the Mediterranean, where I've seen immense stools that must be more than a thousand years old. In Renaissance Italy there was a local variation on the beech-equals-energy equation. The tree was widely cropped in the north of the country to provide specialist wood for galley oars. The mountain beechwoods in Liguria were meant to supply this wood for the Republic of Genoa's navy. But the locals had a more basic use for them, not entirely compatible with their patriotic obligations. In a practice called *ronco* they felled the trees, burnt the branches on the spot, sowed cereals between the stools, and then allowed the trees to grow up again – one of the planet's more bizarre crop rotations. There are still huge living stools in the beechwoods here, relics of those working beeches. Even the modern Italian 'man of the woods', Mauro Corona, whose best-selling *Le Voci del Bosco* takes an almost mystical view of the 'society of trees', regards the beeches, '*di faggi*', as '*gli operai del bosco*', the manual workers of the woods. Something similar to *ronco* was practised in parts of Germany, where hill farming was difficult on the thin soils. *Hackwald* involved sowing arable crops – usually rye, but sometimes wheat or potatoes – in woodland clearings, after the ground had been fertilised with ash from slash-and-burn coppicing.

In southern England, beech-wood was the chief fuel for the Roman glass-works, and continued to be used in this way into the 18th century. In South Wales beech charcoal has been found in an Iron Age hearth near Cardiff. The 10th-century Welsh laws valued a beech tree at 60 pence. Beech timber, unusually, was sometimes used for internal building repairs in Wales, for example at Llangibby Castle in Monmouth, *c.*1286. But, as elsewhere, the tree was chiefly cut for domestic fuel, and to make charcoal for the burgeoning lead- and iron-smelting industries. A series of 16th-century Welsh poems laments the felling of the woods for the English iron-masters. Thomas ap Ieuan ap Rhys mourned a wood in Glamorgan:

> there is no shelter in Coed Mwstwr
> nor firewood because of the iron-works

and goes on to list the loss to local communities of building timber, wild songbirds, badgers, hazel nuts and acorns for pigs, and lovers' trysting places. This was an understandable elegy for a wood of maiden trees, perhaps never cut before, heightened by a loathing of English colonialism. But if the tall trees vanished, the wood almost certainly survived. What the Americans call 'old-growth' forest had simply been converted to coppice. And the South Wales woods continued to sustain the expanding local smelting works until the arrival of coal. That they were using chiefly coppiced wood – beech and oak especially – is confirmed by the accounts of local iron-works. In mid-16th-century Glamorgan, cut wood was accounted for in 'cords', a measure of stacked coppice or branchwood. A cord was 2.4 metres long, 1.2 metres wide and 1.2 metres high. Four cords would make one load of charcoal, and five loads of charcoal would make one ton of iron. The annual cut to supply one furnace amounted to 9,611 cords – over 28,000 cubic metres of small wood.

A group of remarkable mountain beechwoods, all worked in the past, survives in the area round the South Wales coalfield. There are isolated clumps of trees at the foot of cliffs and gnarled coppice-stools in steep hillside woods. In places they grow at a height of up to nearly 500 metres, which makes them the highest native beechwoods in Britain, and the most north-westerly in Europe. I walked up through the woods on the Blorenge, a humped sandstone mountain 6 miles east of Ebbw Vale, one February morning. It had been snowing all night, and the flurries in Coed-y-Person – 'Person's Wood' – made the trees look like wraiths. The trunks were amorphous, caked with blown snow. I could hardly tell one species from another. But as I trekked up the slope, wading to individual trees through what was now about 20 centimetres of snow, I could see that Person's Wood was a broad church, hydra-headed. There were ash and wych elm pollards, tall forest beeches, high-cut stools of oak, patches of underwood with all these three trees, probably last cut at the beginning of the 20th century. Along the northern edge there were lines of polled beeches, the tops of their filigree crowns invisible in the blizzard, as if they'd turned to snow themselves.

I tacked south up the Blorenge to the Punchbowl, a wood lying at about 400 metres. All the way I could see the field boundaries marked by great beeches in sinuous lines, as if the fields' shapes had been determined by the trees, not the trees by the field. The Punchbowl swum into view, a whirlpool of wood with, so my map said, a small artificial lake at its centre. I didn't dare to go and look for myself. The snow had stopped, but the ground was treacherous underneath, steep and rock-strewn and rutted. I kept sinking up to my knees in invisible hollows. But even the edges of the wood were a sight to see. Shoals of 300-year-old beech pollards, vast ashes, immense gargoyles of beech welded to the banks of a sunken trackway, and spouting almost horizontal sheaves of branches over the path. From the highest point of the wood, I could look out over the whole prospect, a landscape simplified by the snow to its basic elements of stone, space and tree. The fuzzy coppice of Coed-y-Person, less than a mile away to the north. The cryptic field boundaries, woods stretched thin. The swirling muddle of the Punchbowl. Much of this whole mountain landscape must once have been wooded commonland, the pastures grazed and the trees cut for fuel.

In the Middle Ages, areas where formal coppicing was carried out were not always as rigidly compartmented as they were to become in more wood-starved times. If there was plenty of woodland around, much of the cutting would have been haphazard – trees harvested as and where they were needed, others lopped of branches or decapitated at shoulder-height. If they have enough sunlight, trees treated in this way respond in the same way as those cut for coppice. They put out sheaves of new shoots around the point of the cut. When the cutting's done deliberately, again on some kind of regular cycle, it's known as pollarding. If the lopping and subsequent regrowth are high enough, the pollarded trees will be out of the range of browsing animals, which means that the production of wood and the raising of cattle can coexist on the same piece of land.

The form of the pollard is even more culturally familiar than the coppiced tree. It's the shaving-brush limes of city streets, the humanoid heads-on-trunks of Arthur Rackham's etchings and Tolkien's Ents. Pollards of most species, lopped on a rotation of about fifteen to twenty years at about 2 to 3 metres above ground, all develop the same kind of crowns, a tuft of new branches emerging from a sinuous and slowly enlarging mass of bosses and scars. But beeches easily win the contest for imaginative rebuilding, generating extraordinary rococo ornaments every time they grow a branch back.

In heavily wooded areas of Europe, areas of pollards in grassland – 'wood-pasture' – were either intermingled with coppice, or coexisted alongside it, with temporary fencing protecting the young coppice regrowth. Terms like *wald* and *weald* (German), *bos, bosch* and *holt* (Dutch), *bosc* (Old French) – many of which found their way into the English language – seemed to be generic descriptions of uncultivated and partly wooded areas, regardless of how the trees were managed and used. The distinction was with the cultivated *feld* or field. Only later did these terms acquire qualifying elements to distinguish open wood-pasture from enclosed coppice-woods. This early open woodland is probably best described by the term savannah, a Spanish word first applied to the open plains of North America, then the lightly tree'd plains of Africa and now widely used for any grazing land with scattered trees and groves, whether shaped by humans or of natural origins.

The medieval Chilterns were probably this kind of mosaic. From the Thames in the south to Wendover in the north, there were 200 square miles of hill country, half of it – chiefly the infertile steeper slopes and the plateaux – covered with mixtures of coppice, wood-pasture commons, and a few enclosed timber lots. This was twice the general proportion of wood in England at the time. Pollen evidence and early account books suggest the local tree-mix was more varied than today. Merton College, Oxford, which owned the manor of Ibstone, recorded its woods as a mixture of beech, oak and ash. A lease of Kilridge Wood, near Stonor, mentions 'beeches, ashes, withes, maples, appsis [aspen] and whitebemys'. Beech was common across the whole region – though not often in tall, uniform stands. In 1310, there was an order to cut down 3,000 beeches in Bledlow Wood. Further south, down the escarpment, Sir William Stonor sold 500 beeches 'to be taken within all the grounde of Saunders and a grounde called Herryes landes at Greenfield'.

The evidence of account books and charters provides even more specialised snapshots than pollen deposits. These records present woods at the points of survey, the decisive moments of cutting and selling. David Roden, and Leslie Hepple and Alison Doggett, who have done exhaustive trawls of the Chiltern archives, record the selective perspectives of men who had a stake in the woods, who were ambitious or plagued by debts, who knew precisely what trees they needed and what they were worth. In this early period they reveal little about the people who worked in the woods and what they felt about the trees. No anonymous woodland poetry has been discovered, nor even early descriptions of the character of the woods.

But on the wood-pasture commons, the framing of the rights of commoners begins to colour in some of the human detail. The lord of the manor usually had the right to fell timber, and owned the trunks of pollards. The commoners had the right to lop the branches and collect dead wood for fuel and house repairs ('firebote' and 'housebote'). Occasionally there were regulations governing the time or quantity of gathering. In the common woods of Shirburn near

Watlington there was one specific week in which fuelwood could be collected. In parishes on the Chiltern escarpment, rights to the common woods were known as 'hillwork'. An inquisition post-mortem in 1293 for Princes Risborough records: 'Two foreign woods [from *forinescus*, meaning woods outside the manorial park] wherein all the men of four townships have free common, and all the free tenants of the manor have their pigs quit of pannage . . . and all the tenants hold a wood called "Le Hellewrk" in common . . .' Pannage, the right to let pigs rootle for beechmast and acorns, was often something the commoners had to pay fees for, though grazing cattle and sheep on the common was usually free.

Bills of sale for cut wood provide more clues about the ends of the beech supply chain and the destinations and livelihoods involved. In 1482, sales from the Stonor estate include wood for 'my master's nailer' [nail-maker], and beech-wood for a shoemaker to make lasts. Herry Parvin 'toke 11 load coppice in Bonnell Hill to his fyre'. Much greater loads of beech-wood were carted down from the south-western woods to the Thames at Marlow, and taken by barge to London. In 1677 John Taylor of Henley Park agreed to supply 1,000 loads of firewood to William Hawkins, a 'woodmonger' of Westminster, for £540, with delivery to the Thames wharf at Greenlane.

Did any beech from Hardings Wood ever make the long journey by cart down to the Thames, and then along the river to Southwark – an expansive prelude to our own introverted tinkerings, backpacking beech logs to the village? The whole region's fuelwood trade prospered because of its privileged river access to London. By the end of the 17th century even Samuel Pepys was remarking on it. A page in his notebooks from 1688 is entitled 'Notes about firewood taken at Henley': 'my enquiries thereon relate more p'ticularly to beach woode, which is said to burn sooner, clearer, freer from sparkle, and to make better coale. yt will keep fire longer that those of oake, though oake last longer in ye burning then beach, the measure and price being (as I think they told me) ye same, or near it.' Fifty years later, Daniel Defoe, visiting Marlow on his *Tour of Great Britain* (1725), gives the first thorough account of the range of uses and importance of beech:

> Here is also brought down a vast quantity of beech wood, which grows in the woods of Buckinghamshire more plentifully than in any other part of England. This is the most useful wood, from some uses, that grows, and without which, the city of London would be put to more difficulty, than of any thing of its kind in the nation.
>
> 1. For fellies [wooden wheel-rims] for the great carrs, as they are called, which ply in London streets for carrying mechandizes, and for cole-carts, dust-carts, and such like sort of voiture, which are not, by the city laws, allowed to draw with shod wheels tyr'd with iron.
> 2. For billet wood for the king's palaces, and for the plate and flint glass houses, and other such nice purposes.
> 3. Beech quarters for divers uses, particularly chairmakers, and turnery wares. The quantity of this, brought from hence, is almost incredible, and yet so is the country overgrown with beech in those parts, that it is bought very reasonable, nor is there like to any scarcity of it for time to come.

There was a strong local market for beech firewood too, including at least one large-scale industrial consumer in the Henley glass-works. But the one use to which beech was almost never put in the Chilterns was house-building, which was still the prerogative of the more weather-resistant oak.

At the time Defoe was writing, the demands on the beechwoods were indeed incredible, and some of the fellings, or projected fellings, were on a vast scale. In 1721, the valuers Walter Knight and Joseph Collier counted 49,380 beeches in the Upper and Lower Woods at Shirburn, worth, they believed, £3,514 6*s*. But Lord Parker, the potential buyer, didn't trust the valuers, and asked for a second opinion from a local farmer, John Toovey. Toovey's verdict on Knight and Collier was scathing. 'I find they are in the Itrest of the Tranting Ffellows which Buy wood to destroy it' – and laid out his own more diligent surveying technique.

> I have heare sett downe ye method I doe itt in. I have a man of sound Judgement in wood, and we have a gerle, which goes with us, for a sifer [sizer]. She have I scraching Iron and maks marks on ye trees as she goes, and my man goes about 2 poale ofe of her, and valews how many of ffott of Beach their is betweene him and her, and as shee moves, he moves; and I goes ye same distance from him and valews all ye beach wood by ye ffot to him; and soe wee goe oer ye wood . . . we are not infalluable. But cann give As good a Gus to what is in A tree as any body Cann.

Toovey's 'Gus' was more modest but more inclusive than Knight and Collier's. He reckoned the beech, lop and top included, at 7 pence a foot, totalling £2,402.

I'm touched by the image of this thin line of grass-roots foresters, making their slow progress through the beeches. The accomplished village girl up front, scratching her mark on the trees, maybe using a tape to measure their girth at the height of her shoulders; the two calculating men behind, making informed guesses about the volume and value of the trunks, jotting down the result in their Bucks patois. It is as if they were linked by invisible ropes, keeping their fixed distances to ensure that nothing was counted twice or overlooked; that they contained the spaces of Shirburn Woods, got the measure of them.

Their survey was a map of a Chiltern beechwood, at the beginning of a great change. Twenty years later, Shirburn Upper Wood vanished from the map, not because it was cut down – beechwoods had regenerated perfectly well after fellings since at least as far back as the Iron Age – but because its site had been turned into a non-wood. In 1747 John's relation Ric Toovey, who owned most of the land in company with two other men, grubbed out the roots and ploughed up the ground. It was the prelude to a shrewd bit of property speculation. They extended an existing farmhouse, built a thatched barn and stables, hedged the new arable fields and named the whole estate Portobello Farm, after a British naval victory. Then they rented it out to a Daniel West at 10 shillings an acre.

All over the Chilterns, and much of southern England, woods suffered the same fate, including one-third of Hardings Wood. The value of arable land was rising faster than woodland because of high corn prices. The easy availability of coal in London meant that the beech firewood trade was losing its most important market. Landowners found it increasingly difficult to justify hanging on to their woodlands. Anthony Mansfield has estimated that between 1600 and 1800, the southern Chilterns woodland cover shrunk from a half to one-third of its total area. Yet many of the ancient beechwoods survived, and still do. Parts of Shirburn Lower Wood are thriving, as are the great woods (amongst those mentioned above) at Bledlow, Greenfield and Henley.

As great a change as the extent of the beechwoods was the change in their character. As they shrank they became paradoxically more beechy. Many of the oaks had been taken out during spells of intense ship- and house-building, paving the way for the pure beechwoods of the 19th century. As the local agricultural writer William Ellis commented in the 1730s: 'It generally happens in our Chiltern, that where a wood of oak has been felled a wood of beech has spontaneously succeeded; but when this once got dominion, it will always be sure to remain master.'

II

The woods in the New Forest must have been beginning to look as beechy as this by the 18th century, too. There is a long-term tendency, much longer than the cyclic outshading of other trees by beech, for forests to drift from a prehistoric preponderance of lime, to a dominance by oak in the medieval period, and then to beech. No one is sure why, though it may be due to grazing, or a very slow depletion of the mineral nutrients in the soil. The New Forest shifted even further towards beech when it was stripped of much of its mature oak during the naval crises of the late 17th and 18th centuries.

Looking out from Lyndhurst today, across the heath towards Matley Wood, it's hard to make out anything but beech trees against the dark holly underwood. I'm here in mid-November, and in the morning light the chrome and orange dazzle of their leaves makes every other tinge insignificant. The trees look as if they've been burnished. There's a slight ground frost, dusting the fading heather flowers below. When autumns were earlier, the heather must have still been in bloom when the beech leaves were gilding, a colour scheme barely imaginable today. I notice, unsettlingly, that there are Samaritans posters in the car parks: odd to think of this exuberant landscape attracting melancholics.

I'd last been in the Forest more than ten years before, brushing up on pollards for a piece of writing. But I'd always been a regular visitor, and an impulsive one too, calling in whenever I passing through the south country. The rattle of the first cattle grids at Cadnam used to make my spine tingle. It was an overture to a different kind of England, unfenced and gloriously unkempt. I'd first come here on camping trips with my school Scout troop, then on birding holidays, watching Dartford warblers on the gorse and nightjars dust-bathing on the tracks. Even as an adult who spent too much time rushing about, I always made the time for stop-offs at Picket Post for a sip of that heart-stopping view east – 10 miles of unenclosed forest and heathland all the way to the oil refineries on the Solent.

But, just as in the Chilterns at this time, I scarcely noticed the trees. They were simply the fixtures and fittings of another stage-set, an environment, not a living community. When they did swim into view for me, it was as part of a revelation about the whole Forest, an astonishment that such a place could have survived at all on the industrialised south coast of England. I imagined it had scarcely changed from how it must have been in the 11th century, maybe even the Iron Age. The groves of ancient trees intermingled with plains of heath and bog were like a vision of the savannah that was our species' aboriginal habitat, and the home landscape of the first pastoral settlers in Britain. It was almost certainly a working wood-pasture, and a kind of common, in late prehistoric times. When William the Conqueror appropriated the area as a royal hunting forest in 1087, he may have helped preserve it from cultivation, but he did not create the character of the place.

Yet it wasn't the ancientness that struck me so much as the sense of newness. The renewed Forest. It seemed to be slowly pulsing with life. The woods were visibly fraying at the edges. Windblown willows lay by the sides of streams, sprouting vertical shoots. The heather and bracken were sneaking in where big beeches had fallen, and young oaks poked out of clumps of holly and blackthorn. The whole place seemed to be in subtle vibration, changing partners, shuffling itself.

G. E. Briscoe Eyre, one of the many topographical writers who haunted the Forest in late Victorian times, caught this sense of dynamic tension in one exact word: 'The slopes that connect the moorland with the timbered lowland partake of the vegetation of both, and form a *debatable* land [my emphasis] between them, where descending tongues of heath interpenetrate the advancing wedges of rough woodland.'

But I can't see much debatable land this autumn morning in the new millennium. The grass plains have been shorn by the expanding herds of ponies. That tangled fringe between the woods and the heath looks foreshortened and abrupt. I don't spot a single oak or beech sapling on my walk up to Wood Crates – the curious and unexplained name for one of what are now entitled 'the Ancient and Ornamental Woodlands'. I'm to be taken on a tour by a posse of high-calibre naturalists, recruited by my old friend, the photographer and writer Bob Gibbons, and including Neil Sanderson, a freelance ecologist with a prodigious understanding of organisms most of us don't even notice, and Clive Chatters, chairman of the New Forest National Park Authority. We tramp off through the melting frost, through the fringes of younger birch and oak that surround the core of ancient trees. When the beeches swim into view they are astonishing, a whole clan of breakaways and throwbacks: immense pollards; soaring, uncut pillars; beeches with the uprightness of timber trees but none of their uniformity; half-rotten crags of wood. There are forked beeches, beeches with rippled or reticulated bark, a single whirled beech, with a trunk like a squat corkscrew, as if it had been twisted by a tornado. Perhaps it had, in its impressionable youth. Gales are great shapers of beech. But some of these forms and bark markings are genetic, and crop up regularly in wild populations. So, perhaps, does the tendency to split. In front of us is a cleft beech, a three-storey wood in its own right. There's a beefsteak fungus jutting out like an ox-tongue 2 metres up, a young Douglas fir rooted in the fissure at 5 metres, a holly in a rot-hole 10 metres up. In the Pyrenees I've seen woods regenerating from these aerial seedlings. In some of the tight-packed ancient stands, the old trees don't fall over but rot where they are, collapsing into themselves like concertinas, and gently lowering the saplings with them, rooted in a compost of spongy wood.

The trees aren't bunched up enough here for that to happen often, but most of the young beech seedlings we see are rooted in decaying trunks on the ground, or in the stumps of fallen trees. There may be more nutrients in these sites, or maybe a degree of safety from earth-bound predators: some of the stump-borne saplings we see are rooted more than a metre above the ground. But there are very few young trees between the ages of about five years and fifty, and natural regeneration seems to be on hold. We talk with Clive Chatters about the long cyclical rhythms of the Forest, which are both ecological and social. Ever since William's act of afforestation there's been tension between those concerned about the wood (what the royal hunters called *vert*) and those who have a stake in the open areas of grazing, the commoners. When the Crown's interest switched from hunting to commercial timber-growing in the 16th century, the contest continued, with the woods and the heaths, in Briscoe Eyre's metaphor, in constant 'debate'. At present, the grazing has the upper edge because of an influx of newcomers' grazing animals, which are eating the regenerating tree seedlings away.

It's a delicate and rarely stable balance. We're all wishing we had another Forest connoisseur with us to explain its intricacies, the late Colin Tubbs. Colin was the Nature Conservancy Council's officer for the Forest in the 1970s and 1980s, and one of its great champions. He wrote two seminal books about its complicated ecological and natural history and, with George Peterken, worked out the long cycle by which it renewed itself. In the 1960s and 1970s, it was widely believed that the Forest was no longer regenerating, and was intrinsically incapable of doing so. It was a reflection of the times. The forestry sector was single-mindedly pursuing a timber programme that was the woodland equivalent of factory farming. Native trees were despised as inefficient. Ancient woods were poisoned and overplanted with quick-growing conifers, regardless of the long-term economic effects. As for natural regeneration, young seedlings weren't even recognised as trees unless they were shooting up, year on year, with the density of rye-grass.

Peterken and Tubbs suggested that, in the New Forest, young beeches and oaks had ceased becoming established round about 1970, almost certainly as a consequence of heightened grazing pressure. But they also discovered, from a meticulous study of historical documents and the Forest's trees themselves, that pauses like these were the rule in its long-term regeneration. The surviving giant beeches and oaks, about 6 to 7 metres in girth, originated during the early 17th century, in the well-lit conditions that followed a period of intensive felling. The second generation, largely 150 to 200 years old, is a mixture of beech, oak, birch, hawthorn, rowan, which appeared in gaps in the canopy where old-generation trees had fallen. The most recent generation – largely birch and oak, with very little beech as yet – sprang up intermittently in the woodland glades and margins whenever browsing pressure dropped during the 20th century. These occasional bursts of regeneration are entirely sufficient to renew the Forest tree cover, and it's odd that the simple mathematics of this process is so often overlooked. An acre of old woodland containing, say, thirty 300-year-old trees can be renewed by just ten new seedlings reaching adulthood every hundred years.

In the early 21st century there are plenty of new seedlings appearing, but whether they survive to become established young trees depends on which part of the Forest they are in. The New Forest is only a single entity administratively. As an ecosystem it is hugely diverse, a collection of linked individual woods and heaths. In areas of heavy browsing, regeneration disappears. In others, where the minerals in the soil may have been depleted by centuries of tree growth and harvesting, it may not happen at all, and the ground may need a spell as natural grassland before it progresses to woodland again, another quiet phase in the Forest's cycle of regeneration. Elsewhere, in patches of holly and scrub throughout the Forest, young oaks and beeches are succeeding, and may be en route to becoming a new generation of 'Ancient and Ornamentals' four hundred years hence.

But Colin Tubbs wasn't confident. In 1986, six years before his untimely death from cancer and unaware of what lay in wait for him, he expressed doubts about whether many of the old pollards – mother trees for succeeding generations – would survive beyond his lifetime. The cutting of pollards – legally to provide browse for deer, and illicitly for firewood – was outlawed in the Forest by the 1698 Inclosure Act. Surprisingly for an area with an ancient history of commoning, the enactment seemed to be respected, and Colin's ring-counts on regrown branches revealed virtually no evidence of lopping after 1700. The consequence, as it has always been for pollards, is a gradual disintegration, as they grow more top-heavy and vulnerable to wind. The only way to ensure the survival of pollard trees – though we might not like the look of the results – is to continue pollarding them.

All the while during this earnest seminar on the Forest's interior history, Neil Sanderson has been providing a kind of naturalist's descant, an improvised recital of how the lives of the Forest's wild inhabitants echo the rhythms in the trees. There's little chance at this time of year of seeing any of the Forest's great insect specialities, the rare beetles and flies that live in dead wood, some of which are found nowhere else in Britain. But there are fungi and ground plants in abundance. We walk over the springy carpets of moss that are a dark beechwood's muted alternative to flowers: luxuriant hummocks of the euphoniously named *Hypnum cuppresiforme* and *Plagiothecium undulatum*, and the plumped velvet cushions of *Leucobryum glaucum*, which often break up into smaller mossballs, and fly about the forest in the wind. There are few common names for these subfusc tuftings: they're the familiars of the dark side of the Forest, too subtle and similar to have entered vernacular language. But there is one that does have a tag, ironically because it is so rare and has such exquisitely finnicky tastes about where it lives. The beech knothole moss, *Zygodon forsteri*, grows only on rain tracks on trunks, and on the scar tissue surrounding water-filled hollows on the exposed roots of ancient beeches, and then only when the trees are growing in ancient native woodland. But there it is, like a tuft of dark green horsehair rug, on the roots of a rotting stump. It causes great excitement. It's a new site for the Forest, and evidence of the ancient beech presence in Wood Crates.

We come close to an area with browsing ponies, and Neil insists that we look, closely. The ponies aren't munching the beech leaves, they're gnawing the bark and buds, working the branches through their mouths as if they were stripping a bone. No wonder they can cause such havoc with young saplings.

But I can't keep my eyes down. The trunks above me look like rolls of translucent parchment. They're covered with a thin and satin-like sheen of lichens. I can recognise a few: the pale grey paint-splash of *Lecidea limitata*, speckled with its black spore tufts; the *Graphis*, or 'writing', lichens, scribbled with hieroglyphics on their thin crusts. But Neil, busy with his lens, knows them all. He shows us one of the New Forest specialities, *Pyrenula nitida*, a 'map' lichen, whose creamy base is broken into a black-edged mosaic like the plan of an early field system. And one that does have an English name, the tree barnacle, *Thelotrema lepadinum*. I look through the glass at the tiny spore-pits; they are exactly like barnacles, not the miniature doughnuts they're sometimes barbarously compared to. There are minute fungi on it. And then, shuffling into view, comes a minute red weevil. It's *grazing* the lichen, just as the ponies browse the beech. Gazing at this long chain of dependency, I wouldn't be surprised if, tucked under the weevil's wings, there were even smaller creatures grazing on it.

Lichens are dual plants, a partnership between a fungus and an alga. The fungus anchors the plant to the tree, and the alga photosynthesises its food. The lichen composites rely on rainwash and bark trickles for their water and other nutrients, and, not surprisingly, are sensitive and slow-growing plants. They cope badly with polluted air or prolonged drought. On big trees, Neil explains, there are clans of lichens adapted to the different micro-climates. Lovers of sunlit wood, and of dappled shade. A 'mire' community, haunting the damp and acid trickles from rot-holes. A 'chalk downland' group thriving on the alkaline sap from bark wounds. Over the centuries the individuals in these communities expand until their edges meet and the bark is covered, and then go into a kind of stasis, a quiet stalemate. They are not dead, just marking time. When conditions change – a rise in humidity, a flood of new light because of the toppling of a neighbouring tree – succession begins again, exactly as it does amongst the host trees. Some species die back, others expand into their territory.

In a group that spans all these communities are those lichens which require stable forest conditions and long-living trees, and find colonisation difficult. The New Forest is rich in these old-forest indicators, and has the finest collection in northern Europe of the lichens of ancient beeches. In Wood Crates, we're in one of their epicentres, a site where the continuity of old trees goes back into antiquity. It was from these focal points that the beech began its expansion through the Forest four centuries ago, and lichens may have migrated to them from the then more widespread oaks. But the beeches here may be representatives of a more ancient lineage still, perhaps the thirtieth generation in the family of pioneer trees that first arrived in Britain six thousand years ago.

We stroll on, past beeches growing, against all the rules, in a bog; through wet glades of bog-myrtle, smelling of balsam in the sunshine; then into darker beechwoods again, full of dead trees studded with bracket fungi. I scratch my initials surreptitiously on the underside of an 'artist's fungus' (*Ganoderma applanatum,* whose spores stain violet when bruised) – an ecologically respectable graffito, I hope. There are stumpy hollies here, too, pollarded specifically to allow more light to the lichens. Holly is the one tree that has never been out of our sight. It's ubiquitous in the Forest, another shade-bearer like the beech, but with the advantage of being evergreen. Its foliage, improbably, is one of the favourite foods of the ponies, and their browse line gives the hollies the air of a low, dark cloister. Smaller bushes are often browsed right to the ground in winter, but they seem to be almost indestructible, and spring back the following year.

One of the features of the Forest are the almost circular thickets of holly, known locally as 'hatts' or 'holms', which develop on the open heath. Traditionally they've been places where young oaks, and occasionally beeches, begin to form new groves, as the seedlings grow up with, and are protected by, the holly. But young holly is as vulnerable as any other tree under current browsing pressures. And new holms, with their cosseted oaklings, are simply not forming at present. Nor are there the numbers of mature oaks there once were, and the Forest as a whole has drifted into a period of beech and holly dominance. It wasn't always so. An exhaustive survey of the trees in 1608, just before the navy began to cast its covetous eyes on them, recorded 123,927 'Timber trees', presumably all oak, and 118,072 loads of 'Fyrewood and Decayed' trees, presumably beech and oak pollards, and representing about 79,000 trees in Tubbs's calculations. A more detailed survey the following year revises these figure down to 122,954 and 53,123 respectively, without clarifying what they signify, but still suggesting an overwhelming preponderance of oak. A survey one hundred years later reveals a dramatically different picture. There are just 12,476 oaks fit for 'Her Majesty's Naval Service' – that is, ready for immediate felling and use. What had happened to reduce the oak population – or at least its older constituents – so much?

III

The 17th and 18th centuries were a time of ceaseless political upheaval in England, and this was reflected in changing perceptions of trees as a resource. There was economic growth at home and military threat from abroad, and the government was torn between short-term exploitation of woods to provide the fuel for industry, long-term conservation of older timber to build warships, and clearing the lot to make land available for growing food for the expanding population. Needless to say, short-term thinking prevailed. During the period of Charles I's 'personal rule', for example, when Parliament was suspended, the Royal Forest of Dean was auctioned off, and no one was surprised that the successful bidder was Sir Robert Winter, unscrupulous entrepreneur and suspiciously rich iron-master. Within a year he'd reputedly felled one-third of all the timber trees. When the king himself was felled during the Civil War, it 'merely substituted', as Simon Schama has put it, 'the spoliation of the many for the spoliation of the few':

> After so many years of being fenced off by contractors, whether parliamentarian or royalist, the woods were simply invaded by great armies of the common people who whacked and hacked at anything they could find. Brushwood, standing timber, fallen limbs and boughs – anything and everything was taken before the next-door neighbour or the next village could get to it.

Beech was cut along with every other kind of burnable wood, but probably benefited overall from the overarching lust for oak.

If this is a fair, if questionable, picture of what happened, it seemed to do little long-term harm. The woods that were looted during the Commonwealth have survived. Even the Forest of Dean, which was comprehensively stripped of its trees, began to recover again after a period when much of it was grassland. In the language of the time most woods were 'wasted' rather than destroyed. It must have been distressing to many observers to see woods of forest trees reduced to coppice in an instant, but in cold ecological terms what had happened was simply a shift in the age-distribution of the growing wood. Large numbers of tall trunks had been replaced by even larger numbers of stubby shoots. Much the same shifts occurred during later national crises. The periods of true woodland destruction, the wholesale changing of woodland to unwooded land, lay centuries ahead, in the middle of the 19th century and the last half of the 20th.

But with the Restoration of the monarchy in 1660, the government decided that enough was enough. The country's strategic reserves of timber were in a critically depleted state, and the Crown Commissioners of the Navy asked the Royal Society to prepare a special report on the problem, with recommendations. Few historians now believe there was a real timber drought, in terms of there not being enough home-grown trees to sustain a navy. The trees were plainly out there, as had been evidenced by the successful building of the fleet that fought the Spanish Armada. Something like 3,000 hundred-year-old oaks were needed to build a single ship. This sounds a huge and potentially devastating harvest. Many of these were the straight trees grown routinely as standards in coppices; the remainder were trees with the forks and quirks needed to provide 'compass' timber for curved features, such as the keel and bow. But oak standards were normally stocked at between 20 to 40 an acre in a conventional coppice wood, so the straight timbers for an entire fleet of 25 ships could, in principle, be found in about 1,000 hectares. That is 16 square kilometres, or just 16 of the squares on a single sheet of a modern Ordnance Survey Landranger map. The compass timber was more problematical, and had to be obtained from free-grown oaks in private parks, royal forests and hedgerows.

The problems in practice seemed to be more about supply than shortage. The Crown Commissioners were short of cash, and private landowners were holding onto their trees to try to raise the price. And both they and the government had good political reasons to talk up the crisis. It was a propaganda blow against the Commonwealth's mismanagement of the country, and diverted attention from the damage caused by Charles I's own casual and self-seeking approach to woodland conservation. The philosophy of Improvement and Enclosure was also beginning to stir, and the hidden agenda of some landowners was a programme of timber establishment that would give them a patriotic excuse for seizing commonland. Political spin is not a modern invention.

Enter John Evelyn, jobbing author, whimsical gardener, and the man chosen by the Royal Society to write what would become the best-known forestry manual in the English language. He could have been born for this commission. He was a sound but discreet Royalist from an affluent, landowning family (his grandfather had built up a fortune making gunpowder). He'd travelled widely, was an intelligent and inquisitive modernist in his ideas, and had experience of growing trees on his own property. Most importantly he was a versatile and popular writer with a journalist's nose for a good turn of phrase. He had written Royalist tracts, books on art, an argument for vegetarianism, and in 1661, a treatise against air pollution: *Fumifugium or the inconveniencie of the aer and smoak of London dissipated together with some remedies humbly proposed by J.E. Esq.* There was an eerie prescience in this early Friend of the Earth's tirade against the 'Hellish and dismal Cloud of SEACOAL' that was poisoning the city, and his inclusion of tree-planting amongst his solutions.

His commissioned report on trees and their renewal – its full title was *Sylva, or A Discourse of Forest-Trees, and the Propagation of Timber in His MAJESTIES Dominions* – was delivered to the Royal Society in October 1662, and published early in 1664. It is diplomatic from the outset, and the reader is assured that the loyal Evelyn would never be so bold as to instruct His Majesty 'in the management of that great and august *Enterprise* of resolving to *Plant* and repair His ample *Forests,* and other *Magazines* of *Timber,* for the benefit of His *Royal Navy,* and the glory of His *Kingdomes*' – though this is precisely what he had been commissioned to do. And once the formalities are out of the way, he launches into his famous first paragraph, firmly linking the health of the state with the condition of its trees: 'there is nothing which seems more fatally to threaten a *Weakning,* if not a *Dissolution* of the strength of this famous and flourishing *Nation,* than the sensible and notorious decay of her *Wooden Walls*'. Much of what follows is in the same waspish style, as Evelyn romps through his autopsy of the state of the nation's trees, their economic and cultural value, and what must be done to restore them. Nothing is too trivial: how to gather the most productive ash-keys, and how to make birch liquor; the beauty of hollies 'blushing with their natural Coral', and the wild service tree's heralding of spring 'by extending his adorned Buds for a peculiar entertainment'. A good deal of the text is derived from earlier writers, and from communications from Evelyn's vast network of correspondents. But his own contributions are practical and sensitive. He is no philistine tree-mechanic, indifferent to the charm and vitality of trees. His view of the beech is mixed. It's 'good only for *shade* and for *fire*', though the wood is useful for making kitchen utensils. But he quotes Virgil's view that the beech is 'sweet, and of all the rest, the most refreshing to the weary shepherd', adding – faint praise perhaps – that the 'very *Buds*', winter-hardened on the twigs, made immaculate toothpicks.

Evelyn's focus, of course, was on the tree that was the very substance of the Wooden Walls, and how it might be replenished: the oak, the nation's symbol, the 'bulwark of her liberty'. He may have been the first writer to use the phrase 'heart of oak' – though he includes it in a paragraph praising acorns as a kind of health food, and stresses that he doesn't mean they gave men *hard* hearts, but '*health* and *strength* and [lives] liv'd naturally, and with things parable and plain'. He gives an exhaustive account of every stage in the raising of the trees, from the gathering of 'mature, ponderous' acorns, to the transplantation of the oaklings into '*sound, black, deep* and *fast* mould'. It's a blueprint for the development of commercial forestry, but full of precocious ecological insight. Despite his devotion to compost, he believed that the densest, strongest oak came from trees grown slowly in rough, stony soils, with the minimum of nourishment. Against the conventional wisdom of the times, he argues that it is not the harvesting of woods for ship-timber or fuel that destroys them, but aggressive arable farming, which turns woods into fields. He wanted to reverse that process, and turn fields back into woods. He acknowledges that nature was perfectly capable of doing this by itself, but that to wait on a '*spontaneous* supply of these decay'd materials . . . would cost (despite the *Inclosure*) some entire *Ages* repose of the plow'.

Deliberate sowing and planting were the only fast, feasible alternatives. It was that ancient dissonance between human time and tree time again.

This, beyond the general fondness for trees he helped nurture, was Evelyn's chief legacy. He popularised the notion of planting woods. It wasn't of course an original idea. Orchards, hedges, small groves of trees, had been planted since classical times. There are a handful of historical records of new plantations in Britain from the 16th century, and an early 17th-century enjoinder to Forest officers 'to caste acornes and ashye keyes into the straglinge and dispersed bushes; which (as experience proveth) will growe up, sheltered by the bushes, unto such perfection as shall yelde times to come good supplie of timber'. By the middle of the century, small-scale seedings and plantings on private land were widespread. But as a philosophy, a policy, woodland creation was regarded as faintly eccentric and probably unnecessary. Why waste energy planting trees, when, since the time of the Creation, woods had done it so very successfully themselves? It was as if Evelyn had included in his *Fumifugium* a recommendation for the building of generators to augment the air supply.

Sylva was a best-seller, and went into several editions. Evelyn believed it had made an instant impact on private landowners, and boasted to the king that it had 'been the sole occasion of furnishing your almost exhausted dominion with more . . . than two millions of timber trees' – a figure he had plucked out of the air, and which he moderated, for the 3rd edition, to an equally arbitrary one million. In fact he seems to have had only a small influence on forestry practice during his own lifetime. From about 1670 there were a few attempts at planting in parts of the New Forest, and in 1698, *an Act for the Increase and Preservation of Timber in the New Forest* introduced the idea of 'rolling inclosure', by which 80 hectares would be set aside each year as a nursery for timber oaks. It was probably the first time in eight thousand years that trees had been deliberately planted there.

But Evelyn's 'heart of oak' symbolism certainly took root. David Garrick published his rollicking sea-shanty – 'Heart of oak are our ships/ heart of oak are our men' – in 1759. John Charnock, a naval historian, floated an even more audacious metaphor:

> It is a striking but well-known fact that the oak of other countries, though lying under precisely the same latitude with Britain, has been invariably found less serviceable than that of the latter, as though Nature herself, were it possible to indulge so romantic an idea, had forbad that the national character of a British ship should be suffered to undergo a species of degradation by being built of materials not indigenous to it.

And in 1763 the Liverpool shipwright Roger Fisher published his report *Heart of Oak: the British Bulwark*. This reiterated Evelyn's worries and recommendations, but had a radical edge in its scourging of the profligacy of the Hanoverian landowning class. It was also, unusually, based on detailed personal research – not in the woods themselves but, more relevantly, in the offices of timber-dealers and shipwrights, where deals were made and wood prices fixed.

It was Fisher's pessimistic conclusions, coming so soon after the big losses of shipping during the Seven Years' War with France, that really sparked planting fever. It became a national contest, with landowners vying not just for big tallies, but for prizes. Colonel Thomas Johnes, Lord Lieutenant of Cardiganshire, won a gold medal from the Royal Society of Arts for planting what he claimed to be 922,000 oaks on his estates. His fellow Welshman, Henry Potts of Llanferres in Denbighshire, won a silver in 1821 for planting 528,240 forest trees, chiefly Scots pines and larch. The empty hillsides of

Wales were to be an irresistible target for vast conifer planations for the next 150 years. And they emphasised that another influence was coming to bear on British forestry thinking, beyond patriotism and the pleasure ground's seductive delights: German forest science, *Forstwissenschaft,* which was steering the growing of trees into a mathematically precise process. It was an empirical technology, based on measurements of the best spacing for trees at every stage of their lives, how girth and height could be traded off against each other, which succession of species was most cost-efficient, and how the optimum volume of timber could be obtained within a geometrically measured unit of space. As Henry Lowood has written:

> The German forest became an archetype for imposing on disorderly nature the neatly arranged constructs of science. Witness the forest Cotta [an early German pioneer] chose as an example of his new science: over the decades, his plan transformed a ragged patchwork into a neat chessboard. Practical goals had encouraged mathematical utilitarianism, which seemed, in turn, to promote geometric perfection as the outward sign of the well-managed forest.

The symmetrical forest had entered the roll-call of ideal types.

In terms of the conveyor-belt production of uniform timber, organised plantations worked. But the raising of wood had moved a long way from throwing handfuls of acorns into the bushes and trusting they would eventually push their way through to reach 'such perfection'; and even further from the time when timber merchants searched for suitable trees rather than growing them to order. John Evelyn's book had played a significant part in a sea-change in our cultural attitudes towards trees. The old idea of wood and timber as bounties given by nature, to be picked as and where they grew – the fruit of wild trees – was fading. In its place developed the idea of trees as artefacts, biddable machines for the production of timber, programmed at every stage of their lives from planting to cutting. The fundamental grammar of our relationships with trees changed. Before, 'growing' had been an intransitive verb in the language of woods. Trees grew, and we, in a kind of subordinate clause, took things from them. In the forest-speak of the Enlightenment, 'growing' was a transitive verb. We were the subject and trees the object. We were the *cause* of their existence in particular places on the earth.

And the changes in the processes of growing brought new kinds of landscape into being. Woods where the heights and forms and species of trees had been in continuous variation, where flowers grew in the spring and mushrooms in the autumn, were augmented – and increasingly replaced – by even-aged, single-species grid-lots, devoid of all but the most adaptable of plants. (The planters could have found even an aesthetic argument for their style in *Sylva*: in the glossary, heterogeneous is defined as 'repugnant' and homogeneous as 'agreeable'.) Coniferous species unknown in British woods before, such as Douglas fir and spruce, became increasingly popular, because of their speed of growth.

As for the beech, it profited by default. Useless as timber for ships and houses, increasingly obsolete as fuel, confined as a wild tree to remote hill country and scruffy commons, it was largely ignored, both during the great oak cullings of the 17th and 18th centuries and in the subsequent replantings. There were a few beech plantations, here and there. Another big Welsh landowner, John Maurice Jones, for example, set down 150,000 trees in Denbighshire between 1804 and 1810, though what he planned to do with them goodness knows, as there was no obvious market for the wood at that time. But back in the New Forest, both old pollard beeches and new seedlings benefited from the light let in where the oaks had been felled, and it was from a flush of seedlings about 1620 that the oldest trees in Wood Crates sprang.

It's late afternoon in the Forest. We're walking back to Wood Crates in a state of smug pleasure, full of the satisfaction of vast quantities of exchanged knowledge, and chatting about arcane topics like the singularity of crab trees and the reasons for the late leaf-fall. Clive points out a young oak about 20 metres from the road. It's an upright, clean-trunked cadet, and looks conspicuous amongst the sprawling beech hulks and sheaves of twisting birch. Clive explains that it's called the Normandy Oak, and commemorates the D-day landings. It was picked out by Forest officers as a promising tree which had most probably germinated in 1945. We all smile at the oddity of such an upstanding and seemingly characterless individual being chosen as this raggle-taggle forest's contribution to the nation's memorials. But we have no excuse whatsoever for condescension. Many of the old trees in the Forest will have started out like this. The Normandy Oak has a long way to go, during which it will be subject to the same kind of time-driven weatherings we have been admiring all day in the ancient trees. As for wildness, those spectacular 400-year-old beeches are only still alive because they were continuously hacked about by 17th-century commoners. The Normandy Oak is likely to remain unmolested, and left free to make a bid for whatever counts as dignity and beauty in a tree in the 23rd century.

Chapter Four: 'The Immediate Effect of Wood'

The plantation system's profound effect on our relationships with trees, persuading us that they were our gifts to the earth, not the earth's gift to us.

I

At the height of the new enthusiasm for planting, the second Earl of Caernarvon declared that trees were 'an excrescence of the Earth, provided by God for the payment of debts'. This was a motive rather lower than the high patriotism advocated by Evelyn. The deliberate planting of trees – a symbolic as well as a literal creation of a piece of property in the earth – soon began to acquire meanings beyond the production of useful timber. It became a ritual display of power over the land, a visible demonstration of status, a bid for continuity. John Houghton wrote of the value of conspicuous trees to landowners: 'they make or preserve a grandeur, and cause them to be respected by their poorer neighbours'. In 1669, John Worlidge, author of *Systema Agriculturae*, remarked: 'What can be more pleasant than to have the bounds and limits of your own property preserved and continued from age to age by the testimony of such living and growing witnesses?' These living witnessess were now to be press-ganged into service, and laid out in disciplined belts. By the middle of the 18th century, landscape designers such as Lancelot 'Capability' Brown were using trees both as a way of suggesting an estate's seamless links to the working countryside round about, and as a way of separating it, of emphasising its superior status. Brown's signature feature, the Clump, was his brilliant but anodyne device for achieving these aims simultaneously. The Clump was a forest in miniature, a woodland canapé, suggestive of nature's mysteries when viewed from the terrace, but shielding out any evidence of real wildness beyond. Humphrey Repton's adjustments to landscapes were more modest than Brown's but still intended to secure 'appropriation' of 'that charm which only belongs to ownership, the exclusive right of enjoyment, with the power of refusing that others should share our pleasure'.

The trees used in these displays were mainly oaks, limes and elms. Exotic conifers were also favoured, not just because of the year-round density of their foliage, but because they added a hint of the breadth of Britain's influence, and its seemingly boundless capacity to import new wonders of creation from across the globe. But the androgynous and socially questionable beech barely gets a nod as an emblem of status. It was still the tree of the poor and the unsettled.

But in the early 19th century William Cobbett set out precise details for the planting of beeches. His book *The Woodlands* (1825), a 'treatise' on the planting and managing of trees, is a curious manual from the often outspoken champion of country people, prefaced by the assumption that 'the inducements to create property by tree-planting are so many and so powerful, that, to the greater part of those who possess the means, little, I hope, needs to be said to urge them to the employing of those means'. No words of encouragement to those who did not possess such means. Cobbett's techniques followed the German style in their precision. The ground was to be measured out exactly into 'lifts . . . 2 rods wide', then an elaborate system of trenches dug across the lifts. The trenches were forked, manured and drained – though in the case of sandy or chalky soils it was necessary to be patient: 'these, if brought up to the top, will be ages before they become fit for planting'. The nursery-grown saplings should be planted small, have their roots pruned and be cut to the ground in the first summer after planting. As to their disposition, Cobbett found nothing more disagreeable than 'a ragged wood, some trees without and some trees with leaves'. He liked 'uniformity of growth and hue', trees of 'the *same size and height*' and the same species. He found it incredible that any man might believe 'a Beech, a Birch, an Ash, an Elm, and an Oak, might all live harmoniously together' despite this being an entirely normal mixture in many of the woods of southern England he must have travelled through. And he poured scorn on the notion that different species had different 'tastes' in soil and nutrients. Cobbett, of course, didn't have the benefit of modern ecology to tell him that different trees do indeed 'feed' on different nutrients. He preferred the bluff farmer's view that, since weeds and corn obviously flourished on the same stuff, so must trees. He was notorious for his curmudgeonly views on natural beauty, and confessed he had 'no idea of picturesque beauty separate from the fertility of the soil'. He viewed trees much as he did cabbages, as utility vegetables whose resilience and adaptability in their natural state was not only irrelevant but a positive nuisance; as raw materials capable of mass production, in which regularity and output were the only considerations. 'It is the height and bulk of the trees,' he asserted, 'it is the quantity of timber . . . that a man ought to look at.'

But at least Cobbett was prepared to wait for his saplings to reach maturity. 'No man in his sober senses', he wrote, would think of transplanting big trees 'upon a large scale, and especially for profit'. Sir Henry Steuart, Bart, would not have ranked high on Cobbett's scale of sobriety. Two years after *The Woodlands,* Steuart published his epic *The Planter's Guide,* or 'A Practical Essay on the Best Method of Giving Immediate Effect to Wood'. He wanted to spread his ideas about how a fully formed wood could be created in an instant 'by the Removal of Large Trees'. The transplantation of individual mature trees was nothing new; it had been common practice amongst wealthy landowners on the continent for some 150 years. But Steuart had big ideas and new techniques. He moved large trees *en masse,* and hoped to create the illusion of ancient landscape with them. The secret, he believed, was a scientific attention to how his chosen trees grew in the wild, especially in the sites they were to be moved from. He advised noting the kind of soil they were in, how slope and prevailing winds and shade had affected the balance of their roots and branches. When they were replanted it was either in the same kind of site, or, better, where their 'weak' sides could be strengthened by light, or by tilting their root-plates. The result was that the trees settled in well, looked natural and conjured up 'The Immediate Effect of Wood'.

His favourite trees were beech, which were enjoying a period of fashionability in his native south-west Scotland. Moving them was a laborious and labour-intensive business. The trees were prepared years in advance by having a circular trench cut round them, which was filled with manure to encourage capillary rooting. When it was time to lift them, men in groups of three worked the roots free of the soil with a 'Tree-Picker', a lightweight, single-pronged pickaxe. 'Every effort', Steuart insisted, 'must be made to preserve the minutest fibres and capillary rootlets entire'. Then the tree was progressively levered out by raising the soil under one half of the roots and then the other, until it could be tilted onto 'the Transplanting Machine'. Steuart favoured the 'best and simplest now known', invented by 'Brown, the celebrated Landscape Gardener'. It consisted of a strong pole mounted on two wheels. Steuart had the Glasgow engraver William Turner make a precise picture of the procession of this machine to the beech's new planting place. It looks like the triumphant parading of a captured gun. The beech, about 9 metres tall and a metre in girth is mounted along the pole. Its roots, earthless save for a ball at the centre, bristle like an immense sea-urchin. They're drawn very accurately, not, as popularly imagined, mirroring the height of the branches, but about 50 centimetres deep, and extended out on the side where the branches are dominant. Two men ('Balancemen') are riding on the tree, three more holding it level with the pole with ropes, and a sixth leading the team of two horses. The journey didn't always go smoothly. Steuart recounts one trip where the ropes holding the root end to the pole gave way. The root stuck the ground with some force, and 'the momentum of the movement . . . pitched the Balancemen (now suddenly lifted to an elevation of nearly 12 yards), like two shuttle-cocks, to many yards' distance, over the heads of the horses and the driver, who stood in amazement at their sudden and aerial flight!'.

No wonder that, after such heroic endeavours, Steuart dedicated his book to the king, 'the Munificent and Liberal Patron of all the Arts' – and the patron, he cannily adds, 'of the Art of Creating Real Landscape in Particular'. 'Real Landscape' joins true woodland amongst our archetypes of natural scenery, as a strident example of the belief that these ideal states could be achieved or re-created or improved on by human intervention. Steuart did not go as far as attempting to replicate the ground flora of 'real' woods, but his scheme included the transplanting of entire shrub layers underneath the forest trees. Whether the result was 'real landscape' is questionable. The plantations he created were simulacra. Not exactly fakes, but facsimiles, lookalikes. For the first few centuries of their lives they would be without much of the plant and animal life adapted to long-established woods. They had no history, no provenance, no 'grain'.

Does this matter? In the 1930s the Marxist critic Walter Benjamin wrote an historic essay called the 'The Work of Art in the Age of Mechanical Reproduction'. He was discussing chiefly the impact of photography on the authority and charisma of original works of art, but his remarks have an uncanny relevance to contrast between spontaneous and artificially created landscapes: 'Even the most perfect reproduction of a work of art', he wrote,

> is lacking in one element: its presence in time and space, its unique existence at the place where it happens to be. The unique existence of the work of art determined the history to which it was subject throughout the time of its existence. This includes the changes which it may have suffered in physical condition over the years . . . One might subsume the eliminated element in the term 'aura' and go on to say: that which withers in the age of mechanical reproduction is the aura of the work of art. This is a symptomatic process whose significance points beyond the realm of art. One might generalise by saying: the technique of reproduction detaches the reproduced object from the domain of tradition.

Woods are not works of art, but mutable living systems, and changes in their physical conditions over time eventually give them the patina of individuality. But replace 'work of art' by 'wood' or 'landscape' in this passage and you'd have a fair description of the way that the mass planting of woods affects our perception of them.

Benjamin did not have much time for aura, which he believed emanated from the private ownership of artworks, or from their being the focus of magical and religious cults. He felt that the making of mechanical reproductions freed works of art from these elitist and undemocratic constrictions, made them truly 'of the people'. Do mechanically reproduced woods, like those engineered by Steuart – and by extension all mass plantations – make the experience of woodiness more available, less tainted by arcane associations with privilege and the past? Or is the sense of *authenticity* of a wood or collection of trees, the 'aura' generated by its ancient provenance or individuality, crucial to our feelings about it? A unique 'presence in time and space' is not something which can be created simply by depositing trees in the ground, however ancient they may happen to be.

Paradoxically, what surfaced simultaneously with the planting movement was a growing sense of trees as individuals. They began to be adopted, measured, granted histories, personalised. In the wild, trees commonly grow in communities – colonies, groves, thickets. In a wood – and especially in a planted, regimented wood – it needs a concentrated kind of attentiveness to separate them. Perhaps the passion for planting helped nurture a stronger sense of possessiveness, of stewardship. A known tree was a solid link with the past, an embodiment of continuity. It could be welcomed as one of the family – or at least the family estate. It could be appropriated as a trophy, a proof of clever husbandry, a symbol of ancient occupation or social standing. When the idea of 'heritage' began to take on its hazy shape, characterful trees slipped easily into the mould, as a species of cultural property.

Named trees, trees marked out, were in themselves nothing new. In Anglo-Saxon charters, specific trees are often listed as boundary markers. During the old ceremony of Perambulation, or Beating the Bounds, they indicated the points where the gospels were read or prayers said. Thorns are the most frequently listed, followed by oak, willow, maple and ash. (Beeches are scarcely mentioned. In the Middle Ages they were still trees almost exclusively of woods, not farmed countryside.) There were curative trees, parish meeting trees, celebrity trees. Trees believed to have been planted by Chaucer, sat under by Queen Elizabeth, hidden in by Charles II.

But all these were trees with a specific function or association. The early 18th-century attitude to the merely ancient or interesting is summed up by the fate of the Greendale Oak, in the Duke of Portland's park at Welbeck in Nottinghamshire. A century earlier, John Evelyn had estimated that 250 cattle could shelter in the shade of its branches. The duchess liked the tree so much that she wrote poems to it – 'As much its height the other trees exceeds, / As they o'ertop the grass and humbler weeds' – and then hacked off the branches to make furniture. The tree eventually succumbed, despite this abundance of affection, and in 1724, the family excavated a roadway through the trunk wide enough for a carriage and four.

This story is told in Jacob Strutt's *Sylva Britannica,* the first true catalogue of 'Forest Trees, Distinguished for their Antiquity, Magnitude or Beauty'. It was published in 1822, complete with the author's own drawings, and subscribed to by an impressive list of the great and the good, including one John Constable, A.R.A. Strutt's introduction is clearly intended to flatter this new sensibility in the landowning class:

> those venerable trees which seem to have stood the lapse of ages – silent witnesses of the successive generations of man, to whose destiny they bear so touching a resemblance, alike in their budding, their prime, and their decay . . . the gratification arising from the sight of a favourite and long-remembered tree, is one enjoyed in common by the peer, whom it reminds, as its branches wave over his head, whilst wandering his hereditary domains, of

> the illustrious ancestors who may have seen it planted; and by the peasant who recalls, as he looks at it on his way to his daily labours, the sports of his infancy round its venerable trunk, and regards it at once as his chronicler and land-mark.

Strutt relied for much of his evidence on antiquarians like H. Rooke, a meticulous measurer of trees, and the more scientifically inquisitive Thomas South of Bath. South's description of the nuances of growth of the Bull Oak in Salcey Forest (11.2 metres in girth at head height in 1763) reveals the germ of an aesthetic delight in the ways trees dealt with the challenges of old age. Writing in 1783 about his earlier observations, he says:

> Its head was as green and vigorous last summer, as it was at that time; and though hollow as a tube, it had increased in its measure, some inches. Upon the whole, this bears every mark of having been a short-stemmed, branchy tree, of the first magnitude; spreading its arms in all directions around it. Its aperture is a small, ill-formed gothic arch, hewn out, or enlarged with an axe, and the bark now curls over the wound – a sure sign that it continues growing: and hence it is evident, that the hollow oaks of enormous size recorded by antiquaries, did not obtain such bulk while sound; for the shell increases when the substance is no more. The blea [the pale, new wood under the bark, now known as the cambium], and the inner bark, receive annual tributes of nutritious particles, from the sap, in its progress to the leaves; and from thence acquire a power of extending the outer bark, and increasing its circumference slowly.

There are only two beeches mentioned by Strutt, both in areas close to his native London. 'The Great Beech' in Windsor Forest reminds him of 'some rude mass of broken architecture' and he remarks that earlier visitors might have imagined they had strayed 'within the precincts of some marauder's cave'. Burnham Beeches, near Stoke Poges in Buckinghamshire, is included chiefly because it is the setting for Gray's *Elegy.* (Gray imagines a villager, perhaps himself, lounging 'at the foot of yonder nodding beech / That wreathes its old fantastic roots so high'.) Strutt's sketch is a scene from the wood, a group of squat pollards with modest mops of young regrowth, and wood-cutters and pigs busy under the trees. It's the only portrait of a working wood in the book. The 18th century's celebration of heritage trees, and of the foresight of 'the illustrious ancestors' who might have planted them, could not yet quite admit that the activities of peasants contributed just as much to their character as 'the lapse of ages' and the benevolent patronage of their owners.

II

These days we take heritage personally. A writer like Ronald Blythe is perfectly understood when he talks about his 'own powerful landscape inheritance'. We all feel an emotional conduit to the minute and particular details of the places that made us, as we saw them. When I moved from the Chilterns to the treeless plains of Norfolk, it wasn't the simple woodiness I missed. It has been a wrench not to be able to slip out of the door and be lost amongst trees, *any* trees, in a matter of minutes, though I can get to a wood of some kind with a 15-minute drive in any direction. What I really pined for were old beeches, towering, bulging, recognisable trees. Waterfalls of wood, like the ones I knew in Frithsden.

East Anglia is not good beech country. The climate, in the interior at least, is too dry. The flat land and light soils have given intensive arable agriculture a bonanza. But there is a loose string of beechwood fragments, almost certainly native, stretching up through the sands and gravels of north Norfolk. So in spring I'm tempted in an unfamiliar direction for beech-hunting, up to the north coast, where at Felbrigg there is the most northerly stand of native beeches in Britain.

The journey north is always a shade poignant, an elegiac trip past the echoes and ghosts of previous woods. In the medieval period the tree-land must have come up almost as far as our house in the Waveney Valley. Just a few hundred metres away Darrow Wood Lane winds north like a snaking forest track, edged by turkey oaks and wisps of hedge, but without a wood in sight. Further on is the Heywood, a 4-mile-long stretch of flat farmland, with one of the densest concentrations of moats in East Anglia. Both names commemorate a considerable tract of wood, probably cleared in the 13th century, though before there were detailed maps to record what had been lost. I drive more slowly than usual, noting the tree species in the hedgerows, trying to dowse what it might have felt like when there were woods all the way. Maple, ash, oak, hazel, dogwood, a few crabs. The odd pollard hornbeam to remind you of what was once the mainstay of the local woods: 'hardbeam' in Norfolk, tough enough to make cogs from; superior fuelwood too, known here and there as 'ay [everlasting] beech', for its habit of retaining its leaves through the winter. But no true beeches, though there is a scatter of prehistoric records for this part of Norfolk. They weren't often tolerated in managed hedges, because of their tendency to shade out their neighbours. Half the hedges along this dead-straight highway are strips of corn-on-the-cob, there to shade and feed the pheasants.

It's another 50 miles before I reach the first beeches. They're on the Cromer Ridge, a line of gravelly moraine, dumped here by the last glacier, that was always too meagre to be worth cultivating. A few old trees in a wood at Thursford, a gigantic pollard marooned in the middle of a beech plantation near Gunthorpe, suggestive coppice-stools in woods around Holt. Coming close to Felbrigg, a couple of miles west of Cromer, the heathy ridge begins to look like a bit of Surrey. There are 'Woodland Hotels' amongst the pines and rhododendrons, arboreal spas. But bright green beech sprays are weaving out between the conifers, and the drive of Felbrigg Hall is edged on one side by a continuous tract of wood.

The hall and grounds have had a Byzantine history, vividly narrated by the last private owner, Robert Wyndham Ketton-Cremer, in *Felbrigg: The Story of a House*. The house was built in the early 17th century, on the site of the old Felbrigg Manor, surrounded by heaths and patches of native woodland. Windhams and Wyndhams from different lines of the family married and intermarried for the next 250 years. They dabbled in politics, went on grand tours, got stung during the South Sea Bubble, built up a sizeable estate. In 1676, William Windham I, a disciple of John Evelyn, began a tree nursery, which he sowed with native tree seed and planted up with local saplings. The 70 beeches came from the nearby woods at Edgefield. In the 1770s, William Windham III gave over the management of the estate to Nathaniel Kent, a well-known 'improver' and a protégé of Robert Marsham, a naturalist and tree-enthusiast who lived in the Norfolk village of Stratton Strawless. Kent believed passionately in the importance of timber in strengthening the finances and grandeur of big estates. In *Hints to Gentlemen of Landed Property* (1775), he argues:

> The true way of managing a timbered estate is, to make use of what Nature has brought to perfection, and to keep up a regular, uniform succession; so at the time we take one egg from the nest, for our own use, we may leave another, as a nest-egg, for the benefit of posterity. Sensible of the importance of this plan, Mr Windham of Felbrigg in Norfolk has done me the honour of

> approving, and adopting it in its full extent; and has impowered me to carry it on upon such a vigorous scale, as will gradually swell the quantity, and value of his timber, notwithstanding his falls will be considerable every year.

Kent may have had faith in nature's ability to bring trees to perfection, but not in its power to guarantee their succession. Twenty years later he wrote with some pride about the plantations he'd established, which extended the estate's woodland far into the previously unenclosed Felbrigg Heath. He explained how the ground had been manured by sheep, how the acorns and beechmast and chestnuts had been sown, how the understorey had been thinned for hurdlewood. The timber trees were already 9 metres tall, and the growth of such 'astonishing floridity' that the whole Felbrigg forest now formed 'a grand bulwark' towards the sea, 2 miles away.

This ambitious expansion came to a halt at the end of the 19th century, during the ownership of 'Mad Windham'. William Frederick Windham was the heir to the Felbrigg fortunes. He was a young man of explosive and obsessive behaviour patterns, shouting in unintelligible noises, going on drinking sprees, haunting brothels, posing variously as a policeman and a railway worker, and 'pointing the attention of ladies to mules while staling'. Today he would doubtless be diagnosed as having a learning disability and probably Tourette's syndrome. But in 1861, aged 20 and on the point of inheriting the estate (his father had died in 1854), he entered into a bizarre marriage with a high-class Piccadilly prostitute, Agnes Willoughby, whose 'protector' happened to be a timber-merchant in his day-job. 'Mahogany Roberts', aka 'Bawdyhouse Roberts', began to cast covetous eyes on the trees at Felbrigg, and schemed to have them included in the marriage settlement. Fearing that not only the honour of the Windham family was at stake but also its earthly estate,

William's guardian (his uncle) took the extreme step of instigating a legal inquiry into his nephew's state of mind. The inquiry was held in open court, and for the next month was a public sensation. Witnesses testified to the young man's outrageous habits and probable lunacy. Others, especially estate employees and local villagers who'd known William since he was a boy, argued that it was all just youthful high spirits. His relative and neighbour Thomas Wyndham Kremer declared: 'I think it perfectly consistent with sanity in a gentleman of rank and property, after coming of age, to dress himself up as a guard and perform the duties of a guard on a railway.' The summing-up for the family – 'if he is to be saved . . . from ruin of health and character, from disease and wretchedness of life' – sounded as much like those earlier pleas for the rescue of our Wooden Walls as the case against a young rake.

In the end the inquiry ruled in favour of William, judging him to be neither madman nor idiot, but just 'a damned young spendthrift and fool'. He was free to continue his wild and profligate life, and by the time of his predictably early death, five years after the trial, he'd bankrupted the estate. It was sold off to a self-made Norwich merchant, John Ketton, whose daughter's later marriage joined the Kettons back with the Wyndham line. In the 20th century, two generations of Ketton-Cremers began to restore the woodlands, planting many more beeches and conifers. Richard Ketton-Cremer was killed in action in Crete in 1941. His brother Wyndham remembered him in the two avenues of mixed hardwood trees in the shape of a V which he planted to commemorate VE day. Wyndham died in 1969, the last of the line, and bequeathed the entire estate to the National Trust.

The avenues still survive, and are now young timber trees. The apex of the V points, intentionally or not, towards the oldest and strangest trees in the Felbrigg Woods. Surrounded by three centuries of planting, but barely affected by it, are clusters of beech trees of quite exceptional age. Most are pollards, including one which measures 7.6 metres in girth, and which may be the biggest (and possibly oldest) beech in Britain. There is also a giant coppice-stool, with four stems, which I measured at 6 metres round. None of these are planted trees. They're remnants of the wood-pasture that existed on Felbrigg Heath before the plantings, and have quite likely lived alongside the Wyndham dynasty for the whole duration of their occupation of this place. Perhaps they understood this, and spared trees of a kind that elsewhere were being treated as so much inflammatory rubbish.

What does a tree need to acquire this heritable quality? As an idea, 'heritage' has the same vague but unquestionable worthiness as 'family values', something to be cherished but not put to the indignity of definition. But what if the family falls out? What is the hierarchy in heritage? When does it *begin*? A few weeks on from my Felbrigg trip, I'm up north again, in another grand landscaped estate. Stradsett Park is in north-west Norfolk, inside the prehistoric range of beech. But no beeches have grown here for the last two thousand years. This is an oak landscape, whose heritage seems to have begun in the early 19th century.

I've come here for an Ancient Tree Forum outing, hosted by the current owner Sir Jeremy Bagge. He'd recently commissioned the landscape historian John Phibbs to advise him on tree-planting, and in the course of his researches Phibbs had discovered that Stradsett was an intact landscape park first created between 1810 and 1813. In the strongroom of the house he'd found a tin box detailing the daily work during the construction of the park. The Bagges had inherited Stradsett, an estate of roughly 1,000 hectares in 1791. T. P. Bagge – another wealthy Norfolk businessman – had ambitions for a park, and paid Humphrey Repton to draw up a design. But it was the up-and-coming landscape architect J. C. Loudon who finally got the job.

There had been no park at Stradsett prior to Loudon's arrival, just a conventional agricultural landscape of arable, pasture and woodland. Onto this apparently undistinguished canvas Loudon imposed a design for an 80-hectare pleasure ground, including a lake, a walled garden and new clumps of oak. Beech was not even considered as a partner to this patrician tree. He used local workers, but employed his own foreman, Alexander McLeish, to keep them in order. McLeish's daily notes survive, recording quantities of earth moved, sums of money spent, and numbers of trees planted. They also recount the destruction, on an almost routine basis, of large numbers of old pollards, which were simply blown out of the ground with explosives. This puts unexpected new dimensions into the seemingly straightforward story of a 'gentleman of rank and property' creating his heritage *de novo*. There were characterful oaks here before Bagge's time, older layers of heritage. And although designers of the Picturesque school were warming towards the ruggedness of pollards, there was still widespread hostility towards them. If planted trees were a visible demonstration of authority and status, so were tall, uncut forest trees. They represented an investment, a stake in the future, an accumulating asset. They also represented an Enlightenment image of elegance. Both values were ruined if a tree was reduced to a pollard. Increasingly pollarding was regarded by landowners as a mutilation – even desecration – of a previously 'natural' object, a 'good' tree. 'Everyone who has the least pretension to taste', wrote Alexander Hunter in his 1776 edition of Evelyn's *Sylva*, 'must always prefer a tree in its natural growth.' There was of course a social agenda behind these tasteful attacks. Pollards were the fuel-trees of the poor, the emblems of commonland. Every landowner knew that, if he was away from his estate for too long, timber trees rapidly began losing their heads. The destruction of pollards was a punishment and a warning, and in that sense a foretaste of the mass destruction of wooded commonland that was to gather pace in the mid-19th century.

Sir Jeremy is directing our gaze across the park. There's not a pollard in sight, but the oaks planted by Loudon, now nearly 200 years old, have turned into fine timber trees. The first exhibit is more pointed, a demonstration that things have moved on from the barbarous practices of the 1800s. We've come to an isolated tree from a period before Loudon's plantings. It's wired up to the sound-wave equivalent of a CAT scanner, the technology of the new generation of concerned arborists. Round the trunk is strung a network of small receivers. When the operator gently strikes a blunt nail against the trunk it sends high-frequency sound waves through the tree, which travel at different rates depending on the nature of the wood in their path. The receptors feed their signals to a computer, which interprets them to build up a map of the varying densities inside the tree, an 'axial tomograph'. Hollows and rot-patches are distinguishable, as are areas of especially dense and strong wood. With this map it's possible, for the first time, to make an informed judgement about a tree's safety and survival prospects and not condemn it at the first whiff of rot. The tree is visible as an architectural whole, weak in some places, braced in others, capable of being declared safe for another generation, or rescuable by the lopping of a single critical branch. 'Ten years ago', the tree-surgeon demonstrator told us, 'we were in the business of condemning trees. Now we are out to save them.'

Gradually a more complicated picture of Stradsett's heritage begins to emerge. Wild trees beginning to be given the same respect as planted. The time-frame of heritage extended, in both directions. Sir Jeremy, a man more than willing to make jokes at his family's expense, conducts us to Loudon's landscaped lake. His grandfather had planted up one of the islands with firs, in the form of a ship. The tallest represent the masts, the shorter the sails. It was a conceit as elaborate as John Wesley's plaited beeches, and just as vulnerable. A colony of cormorants had taken up residence in the firs, and was growing. 'Their guano was sinking the ship,' Sir Jeremy complained, 'and I wanted them killed.' He had written to the Department of the Environment for a licence to cull the colony. The Department replied that he would have to prove 'damage to fisheries'. 'I told them it was not the damned fish I was worried about but my grandfather's ship.' And then the Department trumped him, with either wilful misunderstanding or a rare show of black humour: 'We suggest that if you are so worried about the cormorants you should cut down the trees.'

More revelations about the fragility of the past follow. We have with us John White, connoisseur of ancient trees, who takes us to a dense thicket of vegetation at the far end of the lake. Embedded in it are three immense oak pollards, which John calls the Three Graces. They are the remnants of the wooded common that preceded Loudon's planting. Even by the standards of ancient pollards, they don't look well, and John decodes their history from their shapes and the ring-counts he's made. They sprang about 1500, and grew well for their first two or three centuries. Then Loudon's lake flooded their roots, and his planted trees shaded out their branches. In the last fifty years the pheasant food put out by the Bagges' keepers has nitrified the soil, making it too rich for old oak roots. Sir Jeremy looks rather abashed, and wonders if he should apologise to the assembled veterans, oak and human. But I rather feel for him, sunk as surely as his grandfather's conifers by the internal contradictions of 'heritage'. The heritage of an indigenous bird downs the heritage folly. The heritage lake drowns the heritage trees. Loudon's heritage park destroys a heritage common. When does heritage begin? What are the criteria? What kind of tree qualifies?

III

The combination of whimsy and aspiration that had characterised private planting since the Age of Improvement continued into the 20th century. Beech and other species began to creep onto the agenda but planters were still obsessed with the oak. In 1910, Charles Hurst ('Author of *Valves and Valve Gearing, Hints on Steam Engine Design and Construction* etc.') set out from Manchester with bags of acorns gathered from Sherwood Forest, bound for Devon and, he hoped, immortality. 'If I can leave a track', he dreamed, 'in the form of a noble line of oaks scattered along a portion of my path through life I shall rest content.' As Nelson's Admiral Collingwood had done when on shore-leave, he lovingly buried his acorns in 'retired situations', covering the little pits with leaves and thorns. His bizarre and comic journey is reminiscent of Don Quixote's, as he is continually frustrated and sidetracked. He sits one morning under a cluster of beeches, longing 'to sing the praise of the oak in appropriate verse' and must have have been prompted subconsciously by a memory of the Scots poet Thomas Campbell's excruciating 'The Beech-Tree's Petition' ('Spare, woodman, spare the beechen tree!'):

> Pause, woodman, ere you make a stroke
> Against this unoffending oak:
> Think if there be no other way,
> And let the noble fellow stay.
> But if by hard necessity
> You are compelled to fell the tree
> Then go perform an act of grace,
> And plant another in its place.

'These lines', he continues, 'might be useful, if printed on gummed slips about 6 inches square, and fastened on the trunks of any threatened veterans.' He plants an acorn, as a kind of embryonic, living cairn, on the top of Kinderscout. He cajoles farmers into collaborating with him, by telling fantastical tales of how oak trees are a protection against swine-fever, how they suck up harmful elements from crop fields ('proved by driving a nail into an oak, when a black stain would be seen spreading round the wound'), how they possess the unique property of intensifying the rays of light passing through their leaves. He seals Sherwood acorns in bottles and hurls them into the River Trent, with instructions on how to plant them, and a message from 'an inhabitant of an island in the stormy Atlantic'. He eventually teams up with a stray dog he calls Pontiflunk, and his journey takes on a more free-range character, before ending abruptly in Northampton, after his adopted dog is run over. Hurst almost certainly didn't leave behind him the immemorial line of noble oaks he dreamed of, but *The Book of the English Oak* is a remarkable insight into the planting mind, and one of the most eccentric and rollicking travel books in English since the 'Water Poet' John Taylor's 17th-century broadsheets.

The hero of the French novelist Jean Giono's famous story *The Man Who Planted Trees* (first version published 1954) was more successful. The narrator (Giono himself) first meets the shepherd Elzéard Bouffier, in north-west Provence before the First World War – 'a barren and colourless land', the story-teller believes. 'Nothing grew there but wild lavender.' The next day he goes out with Bouffier, who is carrying a small sack of acorns and a long iron rod. Between turns with his flock, he climbs to the top of the ridge, sticks his rod in the ground and begins planting acorns. 'I asked him if the land belonged to him. He answered no . . . He supposed it was community property, or perhaps belonged to people who cared nothing about it . . .'

Giono next comes across the shepherd after the war. The ten thousand oaks that had survived from the hundred thousand he had planted were now as high as a man. So were the beech trees he had grown from some unspecified source of seed, 'spreading out as far as the eye could reach'. Bouffier's vast plantation, now more than 30 square kilometres, and increasingly stocked with beech, becomes a National Forest Reserve, since the authorities did not want anything 'endangering the growth of this *natural* forest'. The story pits Bouffier the planter, the 'earth-husband', against the destructive forces of bureaucracy and war. At the end of the Second World War he is 87 years old and his trees have become a kind of 'true woodland'. The abandoned villages have been repopulated, the local springs revived. He had caused a 'land of Canaan to spring from the wasteland'.

Giono's story is a touching parable about the contrast between creativity and human destructiveness, an allegorical framing of a proper relationship with nature. It is based on the life of a real Frenchman he met in 1913, but was probably not intended to be read literally. Yet popular beliefs, and then real behaviour, are moulded by such powerful stories. Giono's moral tale of the redemption of Provence is a fantasy, yet suggests a model for practical action. Does it stand up to a real-world ecological scrutiny? Giono doesn't explain, for instance, why there were no trees growing in this empty waste. Had they been destroyed by previous inhabitants, or vanished because of a change in the climate? If so, where did Bouffier find his inexhaustible supply of acorns and beechmast? Where, for that matter, did the locals find the wood to continue their trade as charcoal burners? If the region was, for whatever reason, so inhospitable to trees, why didn't Bouffier's seedlings die? Why weren't they eaten by his sheep, or burned up by drought? Giono suggests that trees can be materialised almost by magic in places where they do not presently exist. His book is a story about tree-planting as symbolic reparation, a way that humans can heal both themselves and the earth.

In the early 1950s, ecologists began meticulously observing the acorn-planting routines of jays. Their vigils were demanding, needing sharp eyes and the suspension of popular assumptions, but they produced revelations about the ecological role of jays in the development of woodland. In Hainault Forest, close to Epping Forest in Essex, M. R. Chettleburgh watched jays not only gathering acorns off the ground but plucking them directly from the trees. They fly away with them in their gullets, or sublingual pouches, and bury them some distance from the parent trees. The areas they choose are almost always open country – grassland, abandoned fields, light scrub, garden edges, big woodland clearings. The purpose is to provide themselves with a larder for the rest of the year, and they bury prodigious quantities. Chettleburgh estimated that over a ten-week period, the birds were making 60 flights in a ten-hour working day, and burying up to five thousand acorns each. A study in Germany found that, over the course of four weeks, 65 jays buried approximately half a million acorns. The autumn total for the whole of Britain may be as high as one and a half billion acorns.

The birds are shrewd planters, too. If a jay has several acorns in its gullet, it buries them all separately, generally between 0.5 and 1 metre apart. The bird pushes the acorn just under the surface with its beak, hammering it on its way if necessary. Then, in a routine which echoes Charles Hurst's little rituals, it covers the hole with leaves and lumps of earth. The jays find the acorns easily, apparently using a grid of nearby vertical features as a three-dimensional map. In spring, they rely on the recognisable first leaves of the emerging oakling as beacons, though a Dutch researcher J. Bossema found that adult jays ignore seedlings grown from acorns they haven't planted themselves.

But the majority are never retrieved, or are pushed back as unsuitable by the jays, and since they've been planted expertly in well-nigh perfect conditions for an aspiring oak, large numbers germinate successfully. These abandoned acorns are the source of the vast majority of young oaks growing outside woods, and the jay is the single most important agency for spreading the species round the countryside. It wouldn't be outrageous to suggest that the oak and jay have evolved as symbiotic partners. The oak provides the jay with one of the most important parts of its diet, and benefits from its acorns being planted in areas where they have a good chance of surviving. And lineages of jays who follow this pattern of behaviour have ensured themselves a continuing supply of acorn-bearing oak trees. The acorn-planting habit is shared by jays and their relatives right across the temperate parts of the northern hemisphere.

Alas, jays don't move and bury beechmast. Nuthatches and mice do, but not in great quantities, and the mystery of how beech moved north after the Ice Age, spread to England and then dispersed itself throughout the southern counties remains. I've never seen beech seedlings in any numbers more than about 100 metres outside an existing beechwood – the kind of modest distance a bunch of mast might be swept in a good gale. A leap of this kind, made once every forty years – the age at which a beech starts producing seed – means that colonising beech moves at 1 kilometre every four hundred years. It ought to have taken the species at least ten thousand years to cross the Channel land-bridge in post-glacial days. Instead it apparently took no more than a thousand. Something was going on that we have no notion of.

Hurst and Giono's shepherd were human jays, symbionts themselves, scattering beechmast and acorns about the countryside for their own pleasure and the general good. You would need to have a peculiarly pure view of 'naturalness' to see their haphazard pokings and plantings as, in principle, any different from the birds'. But the organised 'amenity' planting of recent years is a more problematic business, both in its practice and its intentions. Even the most altruistic of planters have not always had the ecological instincts of jays, or their self-effacement, or their midwife's good sense in letting successful saplings get on with their own lives.

In Britain, tree-planting for non-commercial reasons moved into a new dimension in the 1970s, the decade that saw the beginnings of concern about the environment – and a growing sense of guilt about what we had done to it. The Tree Council was formed at this time, and launched its campaign 'Plant a Tree for '73' – following it with 'Plant Some More in '74'. The tone of these and later planting campaigns was light but the aims profound, going some way beyond simply increasing the nation's tree count. Tree-planting was to become the great ritual of atonement, the way of making painless amends for the devastation our species had wreaked across the planet. Its symbolism was perfect: the penetration of the soil, the implantation of new life, the years of after-care and cosseting. This was a way of repairing the Earth, of nudging it towards renewed vitality, without in any way surrendering our authority over it. Humans knew best. Planting was the environmental equivalent of tough love.

Soon, traces of 18th-century landscapers' agendas – the planting of trees as emblems of status and continuity and good taste – began to attach themselves to the coat-tails of the new planters. Staked saplings began erupting on village greens and roadsides. Commemorative trees were dedicated to retiring councillors, sporting heroes, great anniversaries. On a stretch of ancient chalk grassland just to the west of the Beeches at Frithsden, a commemorative grove of beech and whitebeam was planted by public subscription – just yards from where the local wildlife trust were continually cutting young trees down to keep the downland open. In 1987, in Sherwood Forest, the aluminium industry financed the creation of a 24-hectare Oak Wood Map of Great Britain. The Conservation Foundation, who organised the scheme, wanted every primary school in the country to have 'a tree planted for them in one of the most famous forests in the world by joining in the Aluminium Recycling Can-paign'. It was a bold project, though it didn't quite match the hubris of the 19th-century Highland laird who planted up a hillside with the letters of his estate's name, on a scale that would have made it visible from space.

Conservation bodies are now matching this kind of ambition tree for tree. In 2005, the 200th anniversary of the Battle of Trafalgar, the Woodland Trust launched its Trafalgar Woods project. The aim is to create a 'fleet' of new plantations to commemorate Nelson's victory, which owed so much to the timber from British woods. There will be twenty-seven woods, named after the twenty-seven ships in Nelson's fleet. Britannia Wood in Devon, Dreadnought in County Down, Victory in Kent, Tonnant in Nelson's own county of Norfolk, and Spartiate in the heart of Welsh-speaking Carmarthen – a digression from the ancient and respectful tradition of naming woods after their natural features or the places where they're situated.

But these are trivial worries. It would be absurd – and churlish – to cast doubts on the general value of creating woods as amenities. They transform the look and ecological value of cultivated land and help the planet cope with its carbon emissions. Britain needs vast areas of new tree-land – four times as much woodland as it has now to bring it up to the level of most European countries – and, at one level, a tree is a tree, however it arrives in the landscape. Woodland creation brings communities together, gives children an intimate engagement with living things. In 2004 the Woodland Trust alone involved 50,000 children in the planting of 870,000 trees, an immense adventure in education as well as reforestation. Planting is a powerful signal, too, that a piece of land has been dedicated to woodland. And in some circumstances – in school playgrounds, urban parks, intransigent, over-fertilised grassland – planting is the only way you will see something resembling a wood in your own lifetime, though it will be a wood of little more than standing trees. It takes centuries for forest soils to re-establish themselves on cultivated land, and begin to be suitable for ancient woodland flowers and fungi.

There may be temporal reasons why planting is valuable, too, especially in the case of beech. Few landowners are bothering to grow beech commercially now, with the market flooded with cheap timber from eastern Europe. But it's an increasingly popular tree for amenity plantings in trunk-road verges and field corners – sometimes to the consternation of conservationists, who would like locally indigenous species used instead. But 'local' has become a fuzzy and precarious concept with the advent of global warming. The vegetation of a locality, established over a long period of stable climate, may soon find conditions not to its liking. Its natural response would be to shift towards areas where climatic conditions are suitable. This is scarcely possible – especially for a species like beech whose seeds aren't carried by the wind – in a countryside consisting largely of open fields and criss-crossed by major roads. Nor, given the breakneck pace of climate change, is there enough time for such a natural diaspora. Beech may need a hand to help it colonise new, climatically suitable regions.

But there are other ways of looking at tree-planting. Is it the only, or the best, way of generating woodiness in any particular place? Are its effects limited to the land itself, or do they influence the planters too? Schemes like the Trafalgar Woods project are expressions of the popular belief that woods are human constructions to be imposed on the landscape in the style and manner we choose, and that is a belief that may now need qualification. Somewhere on the long road between the politically motivated invention of plantation forestry in the 17th century, and those children potting up their acorns to save the world, we seem to have forgotten that there is another way in which woods become established. They have reproductive systems. They produce new generations entirely of their own accord, just as they have done in Britain for aeons. If this is no longer part of our cultural memory, it is still a matter of daily experience. Woods of oak and ash and sycamore erupt along railway embankments, and bring trains to a halt with their glutinous leaves. Birches bristle across heathland the moment grazing stops, and invade the gravelly edges of municipal car parks. Even the beech, the finnicky haunter of the shadows, barges its way into the ecological purity of other kinds of woods when it's planted close by. Trees are irrepressible, a fact that conservationists tacitly acknowledge by spending much of their time hacking them down, scything young hawthorns and ash from downland and willow groves from fens. An abandoned wheatfield will turn into a recognisable oakwood in fifty years.

Why don't we take more advantage of this? The benefits of self-sprung trees are huge. They cost nothing. Because they've 'chosen' where to grow, in conditions that suit them, they're more resilient and healthy than arbitrarily imposed trees. They don't need watering or staking or any kind of fussy after-care. Their clumpings and scatter and mixed ages are more natural than the straight and uniform rows that even native trees are usually set down in. But there's that contentious word again. It slips so naturally from the tongue. Is the ancient debate about, so to speak, the nature of naturalness behind our overwhelming preference for planting? Is the tangle of thorn and bramble that young trees grow through – the Chiltern planners' inadequate substitute for 'true woodland', Nan Fairbrother's 'state of original sin in our landscape' – beyond the pale in modern conceptions of the natural? Too wild, too wilful? The brassy energy of successful regeneration may be its own worst enemy. If we don't necessarily want to be in full control of nature any more, we want to help, to be involved. If 'naturalness' entails our physical exclusion from the processes of renewal and growth, a recognition of our irrelevance, then where does this put us in the scheme of things, as we try to rediscover our place in nature?

I've done tree-planting myself, and understand the feeling of gratification when it works. I've bolstered up the tree population in the woodier end of our garden in Norfolk with local specialities – hornbeam and field maple especially – that I felt might not have arrived of their own accord. The saplings are thriving, despite the fact they've been given no stakes or watering or any of the other diligent human care that is supposed to be essential to young trees' survival. But they've been overwhelmed, in numbers at least, by upstart young saplings, spontaneous gatecrashers: drifts of jay-sown oaklings from the big tree in the hedge, cherries from the seeds and suckers of trees already in the garden, ash, sweet chestnut, blackthorn and hawthorn. And, this autumn, a single beech seedling, just 30 centimetres tall, sprang up under the shelter of an oak, and 200 metres from the nearest mast-bearing tree.

I encouraged some planting by children in Hardings Wood, too, even though the place was already stuffed with trees. It's a thrilling and novel experience for them, to be given physical responsibilty for a half-wild organism, to have some influence over its fortunes, to begin to glimpse other kinds of time in the slow and contrary rhythms of the natural world. It can be transforming, too, a chance to have a literal stake in the future. But I wonder what other subtle messages they absorb from their hours of digging and tending. I've just had the Tree Council's package for the 'Seed Gathering Season' 2006. It includes a gathering bag made from brown recycled paper, and the following tips about TLC for newly planted saplings (remember:

'Trees that die are a waste of money'). For *Tending*: 'Check the tree in March or April every year . . . Fill in any gaps in the soil around the roots and use a foot to firm the new soil . . . If the soil is waterlogged, channel/drain excess water away from the tree . . .' For *Pruning*: 'Careful pruning can prevent problems in later life. If a tree has two competing upright shoots, remove one at an early stage to leave a single main shoot. This can save the tree from possible future branch failure . . .' For *Clearing*: 'Pull up any grass and weeds for a radius of at least half a metre around the tree . . . Early in the year, when the soil is moist, cover the cleared area with a mulch mat, bark or brushwood chippings or an old piece of carpet. This . . . reduces competition from weeds, and means that there is no need to use grass-cutting machinery near the tree where it might damage the roots.' As another organisation's leaflet put it: 'A tree is not just for Christmas.'

Care for the natural world is an emotional commitment to be encouraged in children, but it is a treacherous emotion, apt to slip into a sense of custodianship, and then of possessiveness, into a habit of seeing the natural world as not just in need of protection, but unable to thrive without our help. The fostered beechling *becomes* nature, by virtue of our care. What does that make the beeches in the New Forest, still making their own way after eight thousand years?

Natural regeneration is not always simple. It has many new enemies – human, animal, fungal, atmospheric – that were not there in prehistoric times. Its hopeful encouragers need the kind of patience that children haven't always developed. But wouldn't it be challenging for them to become witnesses as well as managers, to observe the processes by which trees renew a wood of their own accord, in ways and associations that we can neither copy, nor always understand?

Chapter Five: 'Feeling Through the Eyes'

Amidst all this industrious forestry, the 18th century discovered natural beauty. Trees, its pundits reckoned, were as pretty as pictures, literally. But underneath their preciousness, had the thinkers of the Picturesque movement sensed something profound about seeing? That it could be a short cut to ecological and scientific wisdom?

I

'THE BEECH', WROTE Gilbert White in *The Natural History of Selborne* (1789), is 'the most lovely of all forest trees, whether we consider its smooth rind or bark, its glossy foliage, or graceful pendulous boughs.' He was one of the first writers since the Welsh bards to use the word 'lovely' in connection with a tree. In all the extravagant tributes heaped on the oak in the 17th and 19th centuries, no one would have dared to use such a cissifying epithet. Even the effusive Evelyn avoided such emotional language, and quoting Virgil's comment on the beech's 'refreshing' nature was as far as he went. In the mid-18th century, the conventional view of beauty was that it resided in humans, or in human creations. Rude nature might have it in embryo, but it needed the art and improvements of humans to reveal it, to raise it to an ideal state. White was in some ways a typical Enlightenment man, but his vision of the natural world, and his intimacy with it, were unprecedented, and they opened up new ways of looking at philosophical issues as well as at wild creatures.

White was the first literary ecologist, the first writer to join together precise and affectionate observations of natural life and landscape with a poetic vision of their significance. And the Hampshire landscape round Selborne where he evolved his new perspective was a place framed by beeches, and dappled by their presence. From his bedroom in the Street, the room where he had his first daily view of the world, you can see the prospect which was the backdrop to his life. Beyond the garden and the meadows, dominating the skyline and shortening the parish's daylight by two hours, is a wall of beeches, a mile-long crest of trees pitching in the wind. It's called the Hanger, from the old English *hangra*, a wood on a steep slope. More beeches line Selborne's ancient green lanes. They cluster on the north side of White's house in a common called Dorton. He lived at the centre of an amphitheatre of beeches. It's no surprise that many of the natural dramas he described were set in beechwoods, or that one of the few sketches of his house he wrote, a nostalgic verse about a green refuge in a beechy vale, was written from a point of view inside the hanging woods:

> Now climb the steep, now drop your eye below;
> Where round the verdurous village orchards blow;
> There, like a picture, lies my lowly seat
> A rural, shelter'd, unobserved retreat.

White was familiar with Dutch landscape painting, and though the English Picturesque movement did not really begin until his last years, he would have understood the idea of regarding a scene in nature as if it were a picture. But he had no trouble holding this view alongside a view from, as it were, *inside* the frame, amongst the teeming detail of the landscape's natural life. On one critical day, the two viewpoints merged, and helped Gilbert make a leap of the imagination which was to prove crucial to his future project, of making a new literature of nature.

The year 1758 had been an uncomfortable one for Gilbert. He was 38 years old, a cleric without a proper living. Though fond of his village and fascinated by natural history, he yearned for a career in the fashionable and intellectual cloisters of Oxford, where he was an absentee Fellow of Oriel College. That autumn his father died, setting off a cascade of financial rumour-mongering that resulted in his being ignominiously rejected in his bid to become Provost of Oriel. He crept back to Selborne to cultivate his garden, and learn how to be content in his 'green retreat'.

A fortnight later he was looking out towards the beechwood on the hill when he witnessed something odd enough to cause an untypical irruption into his pleasant but mundane garden journal. 'Saw a very unusual sight: a large flock of House-Martens playing about between our fields, & the Hanger. I never saw any of the swallow-kind later than the old 10: of October. The Hanger being quite naked of leaves made the sight the more extraordinary.' It was a vignette of the fall turned upside down, of spring birds sprouting where leaves had been, of the season of decline made into one of renewal. Almost from that moment the conundrums of settlement and migration, in his own life and the birds' lives, were to become inextricably linked in his work. He remained in Selborne for the rest of his life, and in a corner of his heart yearned for his favourite birds of 'the swallow-kind' to do the same. White was far from the first writer (or painter) to use the forest as a dramatic framework for great events. But few before had imbued a purely picturesque scene – the martins, summer guests in the village, framed against the winter-bare beeches – with such ecological and personal resonance, or hinted that beauty might sometimes be a narrative quality.

Ironically, the first field naturalist didn't get round to considering the beech as a subject for scientific recording until just a few years before his death. After the publication of *The Natural History* in 1789 he'd received a glowing letter from Robert Marsham, the Norfolk tree enthusiast who numbered Humphrey Repton and Nathaniel Kent amongst his disciples, and who'd occupied much of his life bird-watching and botanising. The two men hit it off at once, and over the next three years they exchanged copious notes – and a good deal of growing mutual respect – about their common passions. They gossiped about what happened to swallows in the winter, about the innocence of nightjars of charges that they sucked the milk from goats, about the unrest in France, about their increasing physical creakiness. But mostly they chatted about trees, and about beeches in particular. In his very first letter Marsham confides that they are 'my favourite Trees as well as your's', and lays out before White the details of his largest specimen. It had grown from a seed planted in 1741, and was now 5 feet 6 inches round and spreading to a circle more than 20 yards in diameter. (He'd encouraged this rapid growth by digging a trench round the perimeter of the root-system.) The glove had been thrown down. White responds with his own best beech, sown in his garden in 1751, and now 4 feet in girth. But he quickly extends the field of play to the wild. Beeches in Hampshire, he insists, grow best on 'steep, sloping grounds, whether they be chalk, or free stone', and in one of these hanging woods there was a tree that rose 50 feet without a bough or fork. Marsham counters with a friend's beech in the

Chilterns, which had 75 feet of clear trunk. The friendly banter continues, the two men working up an aesthetic based on the uncluttered clarity of tall columns like two leek growers at a village vegetable show. But these were the value scales of timber merchants, and White trumps him with a quite different kind of beech: 'the vast bloated, pollard, hollow beeches . . . many of which measured from 20 to 30 feet in circumference! They were the admiration of all strangers.'

I'm intrigued by White's vocabulary here. He used words with a scientist's literalness, not always intending the emotional resonances we now associate with them. By 'bloated' he would have simply meant swollen, without any of the implications of excess or degeneracy that hang about the word today. But the 'admiration of all strangers' is a different kind of praise from his personal declaration of affection for 'the loveliest of all forest trees'. He fell into a similar distancing – as if he were gazing down a microscope – when he wrote about the human oddities in Selborne – lepers, gypsies, an idiot who believed he was a bee . . . Did White see the pollards themselves as strangers, citizens of a different kind of culture from the elegant pillars in the hanging woods? Objects of curiosity rather than love? If so, he was anticipating the kind of arguments that were about to unfold about what constituted 'natural' beauty.

The rivalry between White and Marsham was entirely affectionate, of course. Gilbert was enjoying this late friendship ('Oh, that I had known you forty years ago!') and the discovery of someone genuinely fascinated by the minutiae of Selborne life, and writes to Marsham: 'I begin to look upon You as a Selborne man.' The flattered arborist returns the compliment by dubbing one of his favourite trees Mr White's Beech, and giving him

> some particulars of your Tree. 'Tis about 50 years old, and runs clear to about 25 feet, then about as much in handsome head,

> preserving its stem straight to the top, & spreads a circle of about 50 feet diameter. This I reckon the handsomest proportion for an outside-grove tree . . . for the Lawn or single pasture Tree, I wish the branches should hang so low as only to suffer a man to ride on horseback under them; & the Tree to appear at a little distance like a green hill. These are my proportions for the beauty of Trees in different situations.

Marsham flattered his friend by dedicating to him a beech with such an elegant frame and handsome head. He'd never met White, and can't have known that he was just 5 feet 2 inches tall and as thin as a willow.

Beeches could be manikins, Palladian columns, vegetable trophies. Their attractiveness or fascination could depend on their economic value, their human history, the extent to which they met some ideal image of the fulfilled tree. The idea that the *look* of a tree or a landscape might be an indication of its value – sometimes almost of its moral worth – was beginning to take shape. The word 'picturesque' – 'like or having the elements of a picture' – slipped into the English language during the 18th century (it derives probably from the French *pittoresque*). But the first person to explicitly define the idea of 'picturesque beauty' was another Hampshire parson, William Gilpin, of the village of Boldre in the New Forest. In 1768 he published *An Essay upon Prints: containing remarks on the principles of picturesque beauty*. A decade later, in his fifties, he began the habit of taking sketching trips round Britain. 'His great amusement from childhood was drawing,' he confesses in his third-person autobiography, 'a love of which he inherited from his father and grandfather. And his pleasure on these journeys was to make some remarks on the face of the country in a picturesque light.' It was the Grand Tour done in indigenous miniature. He wrote Tours of the Wye Valley, of the English Lakes, of Scotland. Then in 1791 came his much more substantial and theoretical *Remarks on Forest Scenery* (based largely on his experiences in the New Forest), a book which – hot from the press – is noted by White and Marsham in their correspondence.

Gilpin was no philosopher like Edmund Burke, whose writings on the nature of beauty had dominated early 18th-century thinking. He was an amateur dauber, a summer-holiday sketcher, with an intense curiosity about what it was in a landscape – or a tree – which fitted it for 'capturing' in a picture, and therefore by implication made it pleasing to look at first-hand. It sounds on the surface the worst sort of dilettantism, a reduction of all the layers of association and meaning in landscape to the single index of how they would look with a frame around them – especially as Gilpin included tumbledown cottages and ragamuffin children in his catalogue of picturesque effects. But looked at more closely, his analysis of woodland scenery has, intentionally or not, a sub-text about how trees live in the world. The look of them becomes a register of their relations with each other, with rock and water, with weather and changing light.

Against the conventional taste for symmetry he praises ruggedness and irregularity: 'What is more beautiful, for instance . . . than an old tree with a *hollow trunk*? or with a *dead arm*, a *drooping bough*, or a *dying branch*?' He acknowledges that these ideas 'run counter to utility', but insists that they are a much truer reflection of nature itself. 'These splendid remnants of decaying grandeur speak to the imagination in a stile of eloquence, which the stripling cannot reach: they record the history of some storm, some blast of lightning, or other great event, which transfers its grand ideas to the landscape.' The roots of trees also contribute to their beauty. Old trees, he notes, generally heave their roots above the soil. 'Whether it be a malady or not, it is certainly very picturesque. The more they raise the ground around them, and the greater number of radical knobs they heave up, the firmer they seem to establish their footing upon the earth; and the more dignity they assume.' The pictorial footnote becomes an insight into the structure and endurance of the forest.

He considers the beech in terms of these criteria. He doesn't rate it highly, compared with other timber trees. The branches are too interweaved, run too often in 'long unvaried lines', lack both the firmness of the oak and the 'easy simplicity which pleases in the ash'. The beech's dense foliage gives it the image of 'an overgrown bush . . . It is made up of littlenesses; seldom exhibiting those tufted cups, or hollow dark recesses, which dispart the several grand branches of the more beautiful kinds of trees.' This is the kind of severe painterly dismissal that has given Picturesque ideas a bad name. But when Gilpin considers beeches not as individual figures in a landscape, but in their groupings, he's more generous. Its young generations redeem the tree. They're airy, optimistic. And just occasionally, on poor, dry soils, they preserve 'the lightness of youth, in the maturity of age'. The mature trunk can sometimes achieve this lightness, too, especially when its colour is modified by mosses and lichens. And though beeches often look out of place when they're growing with other trees, together in a 'natural grove' they can be awe-inspiring. Gilpin locates the inspiration for Gothic arches specifically in 'the beechengrove': in its close recesses 'we find this idea the most compleat. The lofty, narrow aile – the pointed arch – the clustered pillar, whose parts separating without violence, diverge gradually to form the fretted roof, find there perhaps their earliest archetype.'

If Gilpin began defining what constituted a picturesque landscape, it was Uvedale Price who gave the concept intellectual backbone. At the start, he entered the debate rather in the manner of a progressive squire (which he was), worried that the curmudgeonliness of his contemporaries was not only holding back the beautification of the land, but unsettling the natives too. He came from a professional, middle-class family who'd taken over a substantial estate at Foxley in Herefordshire in 1679. His grandfather Uvedale Tomkins Price was an accomplished landscape painter, and his father Robert had transformed Foxley, planting large numbers of trees, and creating a model estate, with many small farmers and cottagers and mixed farming regimes. The young Uvedale carried on this tradition. But he was no radical, inspired by the uprisings across the Channel. In a pamphlet, *Thoughts on the Defence of Property*, he had addressed his fellow landowners, and warned them of the rising instability in the local countryside as a result of the Napoleonic Wars and the French landing on the Welsh coast in 1797. He believed the shape and workings of an estate had an effect on public morale and social unity, and that landscapes of the kind designed by Capability Brown – 'drilled for parade . . . like compact bodies of soldiers' – only compounded the alienation between landowner and worker. They conjured up 'the vacancy of solitary grandeur and power'. Price was thinking of more than just the picturesque value of trees when he wrote that 'they alone form a canopy over us, and a varied frame to all other objects which they admit, exclude and group with . . . their beauty is compleat and perfect in itself, while that of every other object absolutely requires their assistance'.

Uvedale was a man in love with trees, and with beech trees especially. Foxley contained 162 hectares of wood-pasture ('open woods and groves') and a further 121 hectares of worked coppice. A good deal of this appears to have been beech, planted either by the Prices, or their predecessors (Herefordshire is beyond the tree's natural range). One of the influences on the development of the Foxley woodlands – and on Uvedale's ideas – was the artist Thomas Gainsborough, who'd become friends with the Price family when he moved from Suffolk to Bath in 1759. The following year he painted

Uvedale's grandfather, posing with some of his own impressive sketches of trees. As a precocious teenager Uvedale kept company with Gainsborough on his rural rides round Bath. He recalled that: 'when we came to village scenes, to groups of children, or to any object of that kind which struck his fancy, I have often remarked in his countenance an expression of particular gentleness and complacency'. Gainsborough's view of the countryside was that of a rather sentimental outsider, nostalgic for the scenes of his boyhood and the honest simplicity of rural life. In his paintings, children, labourers, animals, trees, all have a superficial naturalness, but are often just mood-setting emblems. In his studio in Bath he made three-dimensional models of landscapes, on a table specially reserved for the purpose. He'd use cork or small pieces of coal for foregrounds, mock up a bushy middle-ground from tufts of moss and, more ominously, use spears of broccoli to represent trees. Sir Joshua Reynolds, who disapproved of these manufactured miniatures, noted that they were composed from 'broken stones, dried herbs, and pieces of looking glass, which he magnified and improved into rocks, trees and water'. His mention of magnification is intriguing. He may have meant that Gainsborough was literally viewing the models through a lens, or maybe projecting their shadows onto some kind of screen, turning them into even more impressionist images.

In Suffolk in 1747, when he was only 20 years old, Gainsborough made a copy in chalks of the Dutch painter Jacob van Ruisdael's *La Forêt* (*Wooded Landscape with a Flooded Road*). He captures the wild, tanglewood feel of the original, the people and animals wading through the water, the stumps and pollards, the serpentine anchorages of the roots on the road-bank. But when he used it as a compositional model for one of his own pictures that same year, *Wooded Landscape with Cattle at a Watering Place* (notionally a painting of Drinkstone Park near Ipswich), the details are muted. The forest is thinned. The people are reduced to a single figure dozing on a bank. The half-dead foreground tree is almost identical to van Ruisdael's, but has lost its labyrinthine roots. The lane is unflooded, and leads to a rather gentle and complacent pond. Gainsborough may also have had *La Forêt* in mind when he painted what is the best-known portrait of an English wood, *Cornard Wood, near Sudbury Suffolk,* the following year. It has the familiar structure of groups of trees (obviously oaks) around a central area of open ground with people working and resting (a model which Constable developed later). The trees themselves are straight, tidy, unlopped. Not one is even partially dead. The wood opens to reveal a recognisable church, St Margaret's at Great Henny, in the distance and the picture is often assumed to be from Abbas Hill to the north-west. But no such wood existed in this place in the 18th century. The critic David Blayney Brown has suggested that 'The "unkempt" landscape and the activities of the numerous figures [gathering firewood, grazing cattle] tell us this was common land.' But 18th-century wooded commonland never looked like this, with uncut high forest trees instead of worked pollards. Gainsborough almost certainly painted the picture in London, making a composite image from his softened memories of Suffolk inside the template of van Ruisdael's overall structure.

But he did include individual and characterful trees in his paintings. In the foreground of *Landscape with a Woodcutter and Milkmaid* (1755), there is a dramatically stark pollard studded with knobs and burrs. It's hollow, but sprouting wisps of new twigs. The regrowth and the leaves suggest it is an oak, but the almost barkless trunk has the pale sheen of senescent beech. The milkmaid and the woodcutter, his arms full of sticks, are flirting in the shelter of the hollow trunk. A short way behind them a ploughman works his way across a fenced field. The contrast between these two vignettes – the enclosed labourer ploughing his disciplined furrow, and the dalliance of the two with time on their hands – suggests the differences between agricultural toil and the more leisurely freedom of the commons, of which pollard trees were a defining symbol.

Of course, none of Gainsborough's paintings deserves to be approached in such literal and reductionist way, any more than Jean Giono's story of the man who planted trees does: he was an artist, not an archivist. But his vision of the 'natural' state of trees has helped shape ours. His friend the Reverend Henry Bate Dudley stated that 'nature was his teacher and the woods of Suffolk his academy'. But they were not his photographic studio. His reputation as a 'natural painter' has less to do with exactness of detail than with the open-air feeling of his canvases, the looseness of their structures, the sense of an arcadia where working country folk lived in a benign harmony with nature.

At least this gave him common ground with the benevolent squirearchy of Foxley. In 1760 (the year before Uvedale's father died) he painted a *Study of Beech Trees at Foxley.* The picture is done in brown chalk and water-colour, and has a rather fuzzy look. But the trees are recognisable. At the centre and focus of the picture are two well-spread beeches on a mound, inclined away from each other so that they look like the forkings of a single tree. To their left is an old pollard, recently cut. The curves in the two groupings of trees bring the viewer's eye down to the shaded ground between them, then onto the parish church on the skyline, just visible in the background haze of full-leaved beeches. As Uvedale was to write: 'They alone form a canopy over us, and a frame for all other objects.'

More than thirty years later Uvedale set out the conclusion of his lifetime's contemplation of trees and nature, and the qualities that contributed to their beauty. But his *Essay on the Picturesque* (1794) wasn't some hobbyist's tract, a maverick credo from the redoubts of the Welsh Marches. Arguments about the nature of beauty and its relationship to economics, to the human imagination, even to the social order, were at the centre of philosophical and political debate during the nervous years of the late 18th century. 'Beauty is truth, truth beauty' was still to be uttered by Keats; but thinkers like Price believed that the harmonious scene might define the harmonious human, either as its product or viewer. And that, more urgently, it might provide an antidote to the unsettling affairs in France.

My own copy of the *Essay* is an 1810 edition, published with other texts that indicate the breadth of the debate: Price's correspondence with the landscape designer Humphrey Repton, a dialogue with his friend and intellectual sparring partner Richard Payne Knight, a long-winded and patronising commentary by the editor, Sir Thomas Dick Lauder, spokesman for the Picturesque's pedantic tendency.

The sparring with Knight was ongoing and usually good-natured. He lived a little further up the Welsh border country, in a more shifting landscape, full of rock and rushing water. If Price loved the forms of nature, Knight worshipped its energy. He was sympathetic to the revolution in France, and in 1794 (the same year as Price's *Essay*), published a tempestuous poem entitled *The Landscape*. On the surface this is an assault on the staid landscapes of Capability Brown and William Kent, and the way they suppressed natural forces. But it's also an allegorical attack on the oppression of humanity:

> . . . uncorrupted still, on every side,
> The ancient forest rose in savage pride;
> And in its native dignity displayed
> Each hanging wood and ever verdant glade;
> Where ev'ry shaggy shrub and spreading tree
> Proclaim'd the seat of native liberty . . .

Knight urged the vandalising of Brown's landscapes to release the incipient wildwood. The more restrained Price preferred a stiff letter, though he still lays waste to Brown's whole philosophy, especially his trademark feature:

> the *clump* – a name, which if the first letter were taken away, would most accurately describe its form and effect . . . Natural groups, being formed by trees of different ages and sizes, and at different distances from each other, often too by a mixture of those of the largest size, with thorns, hollies, and others of inferior growth, are full of variety in their outlines . . . But clumps, from the trees being generally of the same age and growth, from their being planted nearly at the same distance in a circular form, and from each tree being equally pressed by his neighbour, are as like each other as so many puddings turned out of one common mould.

If this was part of the argument amongst the landscaping community, there was another with more hard-nosed improvers. The agricultural writer William Marshall wrote a disparaging critique of the *The Landscape* and the *Essay on the Picturesque* in 1795: 'Who but a man totally ignorant of all scenery, except that of a picture gallery, or the wild coppices of the Welsh mountains, could have imagined that woods were in nature, raised with the same facility as they are on canvas . . . by mere *dint of neglect* places, heretofore beautiful, have been rendered picturesk, and highly irritating . . .'

But the Picturesque movement's main philosophical quarrel was with the philosopher Edmund Burke, who in 1757 had theorised that there were two kinds of affective landscapes, the sublime and the beautiful. Sublime landscapes were vast, distant, overwhelming, and generated feelings of awe in the viewer. Beautiful places and things were small, smooth, harmonious, and produced feelings of joy and contentment. Burke didn't exactly see sublimity and beauty as residing in the physical world itself, but as qualities automatically attributed to it by normal observers.

In a couple of decades his theory was obsolete, undermined by its unnatural tidiness, and replaced by the 'associative' theory, championed in aesthetics by Archibald Alison. This saw beauty as existing largely in the eye of the beholder, and argued that the reactions of viewers to different landscapes were shaped by the associations they conjured up. A town-dweller might see a vast forest as sublime or just plain terrifying. Someone who lived there might find it both intimate and beautiful.

'Picturesque' beauty didn't fit comfortably into either theory. The rugged and irregular landscapes championed by Gilpin weren't conventionally beautiful. Nor, except on a large scale, could they possibly fit Burke's definition of sublime. They had an appeal of their own, a sense of antiquity, intimacy, 'naturalness', which might, Price believed, be intrinsic at some level, but which could only be fully appreciated by someone familiar with the theory and practice of landscape painting. His preface is founded on this belief. The *Essay* was to help people 'to judge of the forms, colours, effects, and combinations of visible objects', either by themselves or as components of scenery. But 'such knowledge and judgement comprehend the whole science of improvement with regard to its effect on the eye; and I believe can never be perfectly acquired, unless to the study of natural scenery . . . the improver adds the theory at least of that art, the very essence of which is connection'. One sympathises a little with Marshall's tetchiness after this patronising introduction, and with his view of Price as a mandarin who reduced 'living landscapes' to landscape paintings, who'd never been obliged to 'Gus' the cash value of a tree.

But this is the crux and paradox of the picturesque principle. The view of the landscape and the view of the painting form a closed circle. Perceptions of natural scenery can only be heightened by the study of art. Art can only be refined by more attentive study of natural scenery. Is this the circle of holism or of logic? Landscape painting and the understanding of 'natural' landscapes have been locked in this self-referential bind for more than two centuries, to the extent that paintings may now be our communal models for how scenery – and by implication nature – should look. (Our nagging sense of loss in the countryside is in part a regret that the reality no longer lives up to the fixed images made during one short moment in art.) Is Gainsborough's fictitious Cornard Wood what we imagine 'true woodland' should resemble, and Legozzi's *Madonna of La Verna* a representation of the ideal beech? Price's saving grace may be his stress on the importance of 'connection', which exists independently in both paintings and natural systems.

And in his *Essay* he is soon out in his beechwoods searching for connectivity and talking about qualities that seem to transcend the painterly. At the heart of all picturesque landscapes are

> *variety* – the power of which is independent of beauty, but without which even beauty itself soon ceases to please; the second *intricacy* . . . [which] might be defined, '*that disposition of objects, which, by a partial and uncertain concealment, excites and nourishes curiosity*'. Many persons . . . take little concern in the intricacy of oaks, beeches, and thorns, and may have experienced how differently the passions are moved by an open licentious display of beauties, and by the unguarded disorder which sometimes escapes the care of modesty, and which coquetry so successfully imitates . . .

Later he uses the same metaphor, talking of the 'coquetry of nature', as if it is not the painter or the landscaper producing the effect, but the capriciousness of the wild world itself.

He moves the story into Herefordshire's hollow lanes, full of 'picturesque accidents', and of striking trees shaped, he acknowledges, by the 'indiscriminate hacking of the peasant':

> When opposed to the tameness of the poor pinioned trees – whatever their age – of a gentleman's plantation drawn up straight and even together, there is often a sort of spirit and animation in the manner in which old neglected pollards stretch out their limbs quite across these hollow roads, in every wild and irregular direction; on some, the large knots and protuberances add to the ruggedness of their twisted trunks; in others, the deep hollow of the inside, the mosses on the bark, the rich yellow of the touch-wood, with the blackness of the more decayed substance, afford such variety of tints, of brilliant and mellow lights, with deep and peculiar shades, as the finest timber tree, however beautiful in other respects, with all its health and vigour cannot exhibit.

He laments the simplification, the de-naturing, of a local lane, where a huge beech stretched its roots over a bank: 'the sheep also had made their sidelong paths to this spot, and often lay in the little compartments between the roots'. One day he'd found a gang of labourers covering the whole root-system and its sheep-beds with mould, and laying it as 'smooth from top to bottom as a mason could have done with his trowel'. The tree itself becomes a canvas, drawn over or mutilated by all kinds of forces.

What is it that gives visual fascination to untidy tangles of roots, clean trunks gashed by lightning, sodden mounds of antique moss?

It's possible to understand the appeal of a tall timber beech in terms of Burke's criteria: its majestic height, sometimes beyond our ken; the skin-like smoothness of the bark, the arabesques of its branchings. They conjure the ancient classical ideas of symmetry and elegance. But which bits of our aesthetic, or emotional, consciousness do rot-holes and calluses touch? Price believed that picturesque features were partly a demonstration of the 'spirit and animation' of nature, and partly, like Gilpin, the embodiments of the trees' history and experience: 'Observe the process', he writes, 'by which Time, the great author of such changes, converts a beautiful object into a picturesque one', and then gives an inch-by-inch, year-by-year, account of the possible weathering of a collapsed Grecian temple, which could just as well have been a forest hardwood. Rain, frost, colonising lichens break up the uniformity of the surface. Bits fall off, and birds plant the seeds of berried plants like yew and elder in the chinks. Ivy begins its intricate journey up a trunk and turns it from a bland surface into a catacomb of hollows and ridges and the dwelling spaces of small creatures.

I can imagine Price out in his woods, relishing what he called their 'playful wildness' – but also being playfully wild himself, a small agent of Time's weatherings: lopping a branch to open up a view, shifting a few rocks into the splash of a stream . . . William Wordsworth, who visited Foxley in 1811, chided him for this: 'a man little by little becomes so delicate and fastidious with respect to forms in scenery: where he has a power to exercise a control over them and if they do not exactly please him, in all mood, and every point of view, his power becomes his law'. Foxley 'lacked the relish of humanity' of a 'country left more to itself'.

Wordsworth's strictures make me shamefaced. In Hardings Wood I was delicate and fastidious to a T with respect to the forms of natural scenery, and got to understand very well the seductive licence to control that ownership grants you. I behaved like a matronly gardener at times, clearing brambles round my pet flowers and clipping twigs to give seedling trees more light. The owner of the wood next door was even worse, walking his property in suede bootees with a pair of designer secateurs, pruning cherry branches as he went. Picturesque dilettantes the pair of us. I thought I knew what I was doing at the time, but am not so certain now. In one sense I was entirely selfish. I wanted the wood to be my kind of wood, to my taste. But I wanted it to be ecologically sound, too. I nipped and tucked the vegetation round our primrose clumps, so they would make a better show. I ring-barked the loathsome, spongy, alien poplars, to provide a bit of standing deadwood. But when they started to topple over, years later, I nudged them into a position where they'd take down other trees' branches with them, in a wildwood-circus tumbling act. I lopped sycamores that were shading out ashes, and ashes that were shading beechlings, as if I had certain knowledge of the proper hierarchy in trees. And I remember the excuses I made to myself for gardening in a place that was supposed to be halfway wild. I was simply speeding up its progress towards a more 'natural' state. I was doing no more than would be done by a localised wind, or a tribe of bark-beetles making a corner of the wood commodious for themselves. I was part of nature myself, for heaven's sake, deserving of a niche along with the rest of them.

This was true, but disingenuous. What I – and Price, and everyone who has intervened in the growth of trees for no pressing reasons of survival or economy – what we are all doing is putting our ideal of what a tree should be above its own unfolding – and quite unpredictable – future. And it feels, in the moment of action, not like suppression or control but a release. Robert Marsham releases his planted beeches from the encumberance of low and petty branches to enable them to become what they were destined to be, pure columns soaring to the skies. Uvedale Price protects his pollards (a release from extinction) because they are inspiring demonstrations of nature's resilience against the accidents of history. J. C. Loudon kills his, because in his view they are mutilated beyond release, beyond salvation. He wants the landscape swept clean, born again.

I wonder if the achievement of 'beauty', picturesque or otherwise, is really what this is about. What seems to underlie all these aspirations is a desire to have a parental relationship with woods, to invest them with some kind of order, to put them *right*. It seems to be an almost universal compulsion. When volunteers came to work in Hardings Wood – friends of the trees every one – cutting ragged and rickety trees down was what they wanted to do. I was astonished by the greedy energy with which they stripped and logged the thinnings. It was like a speeded-up film clip of jungle ants reducing a dead animal to a skeleton. On one occasion I gritted my teeth and allowed in a party of young men on a work experience scheme. They set about the place like a gang of manic 18th-century improvers, lopping low branches at will, trimming bushes, straightening things up until I had to step in and ask their supervisor to call it a day.

Is all this compulsive activity just a tic, a reflection of modern tidy-mindedness? Is it a cultural memory of our agricultural forebears, working in their woodland clearings? Might it be something more fundamental still, a genetic echo of our primate ancestors in the savannah, making paths and playing with twigs? Whatever the reasons, we seem bound to be compulsive fidgeters with trees – unless we make a conscious cultural decision to step back from them.

Increasingly Price's *Essay* reads to me less like a mannered appreciation of landscape based on a familiarity with painting than an intuitive insight into how the whole web of natural woodland works, including the humans in it. Price's final tilt at Brown is a vision of the diversity of a wild wood-pasture. 'Natural groups are full of openings and hollows; of trees advancing before, or retiring behind each other – all productive of intricacy, of variety, of deep shadows, and brilliant lights. In walking about them, the form changes at every step; new combinations, new lights and shades, new inlets present themselves at every step.' The point of view is that of the human observer; but it's equally the 'view' of the forest itself, as it falls, opens up, regenerates. Against the mercantile temper of the times, Price set out to write an artistically informed guidebook to what he saw as natural beauty. A tract on diversity, intricacy, accident, weathering, light and shade, fractal surfaces, decaying wood, the superiority of 'the spontaneous trees of the country', and above all the *connectivity* of features in a natural landscape. These are the qualities which he also saw as crucial to a good painting. But they are something else too. This list of the defining characteristics of beautiful scenery is also a fair precis of the criteria by which scientists evaluate the ecological richness of natural systems. Might picturesque beauty be an index of the complexity of nature?

I'm fascinated by this apparent congruence between the judgement of aesthetic philosophers and the way scientists describe and categorise natural landscapes. Are there deeper roots to our emotional responses to the visual than either Burke or the Associationists imagined? What resonances, for example, do 'picturesque' trees call up? They certainly carry associations, of old and 'uncorrupted' landscapes, and of the two hundred years of landscape painting which has featured them. But what attracted the painters to them in the first place? Where did the notion that they were picturesquely beautiful emerge from? What deep-rooted associations do old trees conjure up? Are they some kind of portal to understanding the deep relationship between wildness and time?

II

The autumn colour in Burnham Beeches is flowing down the trees. They have the look of upholsterers' colour charts, the browns at the tips melting to bronzes and then to the gold-flecked greens of the lowest branches. Some sensitive chemical response related to height, or to changes in temperature or sunlight, or the reach of the sap flow is at work. The beeches look busy, flagging up their changes.

Both Gilpin and Price thought that autumn was 'the Painter's season'. The muted colours suggested wistfulness, a time of comforting reflection. The leading characters of Orwell's *Keep the Aspidistra Flying* found rather different associations. Gordon Comstock, the disgruntled would-be poet, and his bashful but ever optimistic sweetheart Rosemary, decide to have a day out in Burnham Beeches to escape from money-mad London and the confinements of their bedsits. They plunge into the woods, Gordon, at least, smouldering with unrequited love. All round them the beech trees soar, 'curiously phallic with their smooth skin-like bark and their flutings at the base'. Released from the city, Gordon and Rosemary are in a riotous and sharp-sensed mood. They fall into ecstatic enthusiasms about everything they see: the colour of a jay's feather, the curving depths of beech boughs reflected in a still pool.

> They discussed for a long time what would be the best epithet to describe a beech-tree. Both agreed that beeches looked more like sentient creatures than other trees. It is because of the smoothness of their bark, probably, and the curious limb-like way in which the boughs sprout from the trunk. Gordon said that the little knobs on the bark were like the nipples of breasts and that the sinuous upper boughs, with their smooth sooty skin, were like the writhing trunks of elephants.

The leaves have just fallen, and Rosemary wades around in them up to her ankles, as in 'a weightless red-gold sea'. 'Oh, Gordon,' she cries, 'look at them with the sun on them. They're like gold.' 'Fairy gold', Gordon mutters. 'As a matter of fact, if you want an exact simile they're just the colour of tomato soup.'

I'm still in a state of puzzled excitement about that odd synchrony between scientific description and picturesque sentiment. I know the cynical Gordon is closer to the truth than the romantic Rosemary, but I don't see that their two truths are necessarily contradictory. At school, I was educated as a scientist, and enthralled by the mechanics of life. But I became uneasy about science when I started writing, fearful that in some way it might 'contaminate' my imagination. I kept it in quarantine in a kind of mental biohazard enclosure, taking it out only under the strictest security. But I'm wiser now, and increasingly believe that finding a common ground between respectful, objective views of nature and respectful views of our own visionary images is what 'nature writing' ought to be concerned with. So I dig deep into the biochemistry I once knew, and swot up on what is really happening in the rites of the fall, hoping that it may reveal something about the beauty of trees in autumn, or at least their meaning.

Autumn isn't a season of decline. It's a time of furious activity by trees, the opposite of the slow windings-down of senility or hibernation. Leaves are probably shed to help hardwood trees reduce water loss during the winter months, when cold ground water doesn't easily enter the root cells. But leaf-fall also provides an opportunity for the tree to get rid of waste products built up over the year, including toxins absorbed from the soil. In some cases, the levels of poisonous metals in leaves increases a thousandfold just before they're shed. At the same time the tree is breaking down the chlorophyll and sugars in its leaves and withdrawing them into its woody parts, conserving them. When the green goes, what is left are the brightly coloured carotinoids – orange and brown and yellow anti-oxidant chemicals similar to those that make tomatoes red – which are believed to bind with the toxins. This flurry of chemical activity is stressful for the leaves, and to protect them during the crucial transfer of chlorophyll many trees synthesise yet another anti-oxidant, the bright red anthocyanin. The final mix of all these pigments, and the tints of specific trees in a particular autumn, depend on factors such as summer sunshine and rain, soil drainage, early frosts.

That is what the season of mellow fruitfulness is all about. The time of high colouring isn't a signal of fading away, but of detox vitality, ruddiness, rude health. Gordon's intended rib-tickler about a summer-vegetable soup actually echoes Keats's upbeat view of autumn as the 'Close bosom-friend of the maturing sun'.

I've been to Burnham before, and every time it astounds me, just as it did Gordon and Rosemary – such an improbable collection of 'sentient' trees only 25 miles from a capital city. In the late 18th century its vast tree-lapped spaces must have been one of the most desirable building plots in the country. But it was its closeness to London that, paradoxically, saved it. The entire wooded common of Burnham came up for sale, with development a distinct possibility, in 1878. This was just the time when the value of green spaces for urban populations was at last being understood, and the Corporation of London stepped in and bought the whole area.

Burnham Beeches' history before that is vague. Francis George Heath, who initiated the campaign to save the wood, believed that the beech trees had been pollarded 'at some remote period of their history' – and certainly not by the local commoners, who did not have the right of lop and top. A more recent local history, by A. D. C. Le Sueur, suggests, without presenting any evidence, that Burnham was an oakwood five or six hundred years ago, with just a few beeches, and that this was the period when pollarding commenced. The first suggestion is likely, the second barely credible, given the 4,000-year-old history of the practice. In the absence of real evidence, myths have proliferated. Cromwell's soldiers did the lopping, to make musket-stocks. Or perhaps the Duchess of Monmouth, in memory of her husband, 'lately pollarded', as Peter Marren delicately puts it, 'by his uncle, King James II'. Any agency, it seems, but the local people, gathering fuel, though the sketch Jacob Strutt made in 1822 clearly shows pollards which have been cut by some hands in the past twenty years. Whatever their origins, the existing pollards – many hundreds of them – started life about four or five hundred years ago, and still look full of life.

But what kind of life? What is it about the frozen, rumpled energy of these grotesque trees that touches us? I want to do an experiment, to stand close to them and try to look with picturesque and scientific eyes simultaneously. I have a sketch-book and a pair of binoculars, which I hope are the right sort of tools.

Burnham is a distracting place, not an easy setting for intense concentration. There's a slight mist, muting the horizon. Drifts of trees seem to billow over the gentle rise and fall of the ground: tall forest beeches, dark tangles of holly, ancient oaks, birch groves, hornbeam stubs, plains of pollards, fallen trees, logs. Burnham is a *whole* beechwood, with all the beech's natural companions, and trees of every age from seedling to rotting hulk. I come across little huddles of planted beeches, marked with plates commemorating the spade-work of aldermen and mayors, the grandees of the Corporation of London. And, only feet from them, sheaves of naturally regenerated beechlings. Odd, this municipal compulsion to lord it over the young natives. But better, I suppose, than clamping impertinent plaques of provenance to them.

Burnham's stewards do well by their trees. I come up against a diversion in the path. An oak has lost a huge branch, and looks marginally unstable. There's a makeshift brushwood fence around it, and an apologetic notice, explaining that if the public would be considerate enough to keep away, the oak can be left as it is, with all its character and cast of attendant creatures intact, and not be submitted to further mangling for the sake of the Health and Safety Executive. Everywhere jays are ferrying acorns about. Red kites float above the trees. I glimpse a sparrowhawk coasting the crests of the beeches, the tension in its glide seeming to create a bubble of stillness over the leaves. And, just for a moment, I have an entire ecosystem framed in my gaze – the hawk over the ripening leaves, the planting jays, the oaklings studded with marble-galls, the forest floor alive with beetles and toadstools.

I try to make sense of the fungi, racking my memory for the knowledge I had thirty years before. There are shoals of miniature puffballs, rippling over leaf litter and rotting wood, gelatinous beech-tufts drooping from dead trunks, tawny grisettes, with the rims of their bronze caps edged with grooves, as if they'd been delicately scored with a knife. But it's the year of the false death cap, *Amanita citrina*, a lover of beechwoods, and deserving a more appetising name for its cup-shaped cap, flecked with lemon rind.

I reach the first of the pollards, Burnham's great legacy, part of what the place was saved for. Even by picturesque standards it would be perverse to call it beautiful. The entire top of the tree must have been blown off years ago. The old trunk is now topped with a snake's nest of coiled scar-tissue and a huge round boss. A dark hole where a major branch has been shed is full of rotten wood, and a young holly has rooted itself in the mould. Out of this crumbling woody boulder, a brand new trunk about 30 centimetres across has climbed up another 6 metres, and produced a respectable head of leaves. And a mast crop. Not a beautiful tree, but a triumphant one. When I try to sketch it, leaving out all the intricate twiggery, it looks like a cave, a dark centre surrounded by craggy outcrops.

Fifty metres on, another pollard, another story. A broad-shouldered tree, its two main branches raised up like a body-builder showing off his biceps. Along each limb rashes of pimples where smaller branches have been sliced off. Or so I imagine. I'm thinking pollarding automatically now. But small raised blisters cover the trunk, too. Might they be a genetic feature, like the chequering sometimes seen on beech bark? It's hard to credit that these trees are all of the same species, so inventive are the ways that growing beech-wood responds to the simple act of chopping into it.

I have a sudden image of my friend Maurice Cockrill's paintings. In 1991 he produced a series called 'Entrances', in which each work began with a cheap, off-the-peg wooden door. Maurice took a felling axe to the doors, creating a random pattern of gashes, and then 'repaired' the wounds by building up restorative layers of paint around them. He once described his technique as 'plastic automatism' – and having seen him at work, his lumberjack's arms furiously following the logic of accumulating paint, I think I know what he means. His work is 'automatism' only because it is more like organic growth than planned development. Eight years later he produced the 'Generation' series. These also have a dark centre, a kind of quarrying, in which a bright nucleus of paint seems to energise the growth of new forms beyond the shadows. Above them tiers of hook-like forms are pulling the growth – 'the regeneration', I innocently called it when I wrote about the paintings at the time – up towards the sky. Maurice's pictures look nothing like pollards. But gazing at the dark hollows in these Burnham trees, and the extraordinary new forms emerging out of them, I'm thinking that pollard beeches grow in the same way as he paints. They mend their own wounds, laying on whorls and braces and arches of woody tissue that are shaped by the topography of what is there already. They have the same chaotic exuberance of form as solidified streams of lava. Or paint. But as I wander past more trees, I begin to glimpse patterns of a kind, common motifs. Fat folds of scar tissue, woody lipoma, so massy it's hard to see how the boles can support them. Branch stumps as smoothly rounded off as a cup turned on a lathe. Thick internal roots snaking down inside hollow trunks. Puffed sleeves round emerging branches. The trunks and branch supports of the pollards all seem to be curling in on themselves, trying to hold the centre.

Then I discover 'the Cage Pollard' where the tree's defence of its centre of gravity is carried to the outer limits. There's a green plaque nearby. In 2002 the queen designated it 'One of the GREAT BRITISH TREES in recognition of its place in the national heritage'. At the foot there's the name of the sponsor: the National Grid. I wonder if they became involved to make an elaborate visual pun. The Cage is a tree whose bottom few feet must have once looked like an enormous barrel, from which a much thinner trunk rises. The barrel hollowed out long ago, and then started to lose parts of its surface, until on one side it's now just an open grid of four barkless staves, a pylon, holding the tree indomitably upright. I'm learning the vocabulary of these beeches, and think I can intuit what's happening, but I need a grammar.

Claus Mattheck is Professor of Biomechanics at the University of Karlsruhe. The biographical note on his books says he likes large-calibre weapons and Staffordshire bull terriers. Not perhaps the kind of man you would want to meet at night in Burnham Beeches. But in reality he's a soft-hearted tree-hugger. I first came across him in a children's book he'd written and illustrated, with the improbable title *Stupsi Explains the Tree: A Hedgehog Teaches the Body Language of Trees.* I thought of this title as I gazed at these wildly extrovert veterans. I've been trying to keep anthropomorphism at bay, but now the expert says it's allowed. He means the tree's body, of course, but the resonance of that phrase won't go away. I knew when I first read it that it was going to help make sense of the way we relate to old trees. Mattheck's academic books are a hard read if you're not familiar with physics; but Stupsi is the best sort of tutor, if you can endure a hedgehog that looks even more like a small human than his tree-men cartoons.

A tree's body language is more extensive than you might expect. It's also easily intelligible, even sympathetic – though the way trees exhibit and then respond to stress is entirely different from most other living things. It's a protracted and permanent expression, frozen into the hard forms of wood. For example: trees growing close to each other, so that their branches rub, often fuse together at the point of friction. The trunks below the branch-weld stay thin, while those above it grow thicker. They've constructed a frame, in which the lower parts of the two trunks are helping each other to support the wind load. Trees in persistent wind grow bent in its direction, but on the windward side everything toughens to keep the tree upright. They grow 'guy-rope roots', buttresses, strong new trunk-wood.

The universal principle which underlies the shaping of trees is the one which defines successful design in all structures, life-forms included. It's called the 'the axiom of uniform stress'. There must be no weak places. Whatever loads or stresses the tree experiences have to be distributed evenly over its whole structure. When a tree is weakened in some way – by disease, lopping, wind damage, lightning strike – it will try to neutralise its weak spots by growing a special kind of tissue called 'reaction wood'. In deciduous trees it's called 'tension wood' because it is a pulling, contractile tissue. The Professor says it is 'white and shimmers like a nylon shirt' (though in the Chilterns wood-turners called it 'sleepy' wood). It functions so like animal muscle that it would be contrary not to think of it as such. Flexed biceps are just short-duration tension wood.

Tension wood grows along the upper side of tilting trunks, and on the joints of big branches with heavy loads. If a branch drops off, a variety of it grows around the weak and exposed area that remains, becoming thick and dense to compensate for the place where there's no wood at all. This is the wood that forms the lips and bosses where branches have been pollarded. If a branch become superfluous, or a burden, the tree creates a tough collar in advance, at precisely the point where it will shed the branch. Bulges in trunks are likely to represent the first stages of hollowing out, as the trunk reinforces its outer layers.

Back in Burnham, I'm looking at a spectacularly tilted beech, a high-wire balancing act. It's sloping away from me at an almost impossible angle, about 40° to the vertical – as far as it could go, I'd say, without collapsing under its own weight. Hard to guess how it got into this position. First tilted in a gale maybe, then slowly sinking as it tried to grow itself back to uprightness. The whole core of the tree is missing, maybe discarded as useless ballast, so that the trunk is like a trough. The rims of the trough are massive tension-wood muscles, hauling it back. There is a twisting mesh of crooked branches at the top end pulling it the other way, down towards the ground, so the tree has responded with flaring root hawsers and a long single branch, both growing against the direction of the tilt. The trunk has become a lever, perfectly balancing weight with muscular tension.

I try it myself. To hell with detachment. I lean forward at the same angle as the tree, imagining my feet pinned down by straps, and trying to pick up a huge weight with my hands. It's a ludicrous posture, and I know it would break my back if I tried it for real. Unless I had tension wood up my spine, doing the pulling.

Burnham is full of humanoid trees like this Weightlifters' Beech. A League of Health and Beauty tree, doing an elegant midriff twist. A Stilt-walkers' tree. A beech with a wooden Zimmer frame. All of them are exercised, like us, with the business of keeping a rather disorderly mass of tissue upright in a turbulent world. You are beyond anthropomorphism in Burnham, into a place of more mutual metaphors.

But a few of the pollards have picked up names because of another kind of human association. Gray's Beech, supposedly the subject of one of the final stanzas of the 'Elegy in a Country Churchyard', went down in the 1930s. The remains of Jenny Lind, on whose roots the 'Swedish Nightingale' used to perch when she was staying at East Burnham Cottage, is surrounded by a safety fence. Mendelssohn's Tree, whose dappled shade is said to have inspired him while he was writing the incidental music for *A Midsummer Night's Dream,* had its top blown off in the gale of January 1990. All the old trees, including these barely living butts, have their own numbers, stamped on small aluminium plates. I like this way of registering their individuality, rather than subsuming their existence under some human's name.

Standing on a small mound, and peering at one of the labels, I can see that I'm at number 01325. It's a conventional pollard, with a decent head of branches. But the trunk's surface is beginning to break up. Some fungal infection is causing flakes of bark to lift up, like scabs, and a thin trickle of sap is running down the tree. And when I look closer I see that the bark is alive with animals. Small spiders are rushing about, in zig-zag exploratory dashes. A stream of wood ants is moving against the sap current. I don't think they're drinking it, but they're collecting minute scaps of bark debris and ferrying them down the tree. I follow the thin line of downward traffic and discover, a shade embarrassingly, that I'm standing on their nest, a vast pile of tiny pieces of wood and leaves. Number 01325 is a very desirable address.

But I can't make any sense of number 01243. It makes me feel uneasy. It isn't a tree from any tradition I know, not picturesque, or noble, or intellectually amusing like Wesley's wooden sermon. It is scarcely a tree at all, just two snail-shells of wood perched on a stalk. Or the skeleton of a prehistoric bird, standing on one leg. Or a voodoo warning. Or an immense fossil embryo. I can't stop these resemblances crowding into my head. How else do you make visual sense of an illegible life-form without comparing it to other living things? But is it living at all? I go close enough to touch the tree, and can read what might have happened. The shells, almost level with my eyes, are part of an immense shoulder of wood half hidden by the foliage of other trees, which itself looks like the remains of an even bigger crown. The embryonic whorls are a tangled turk's-head of tension wood and scar tissue and braces, which the tree has grown to try and keep its balance while its top fell apart. And the bird's leg – a grooved tube of tension wood about 25 centimetres wide – is the final filament of trunk that is holding up the whole extraordinary structure. And it's working. The tree has kept its thin sheaf of branches in the light. Above me I can see its autumn leaves, with not a sign of 'the condition of beech'. It's covered in mast. And all around, where the collapsed crown has opened a space in the canopy, there's a forest of seedlings. I guess that in fifty years they will have shaded their parent to death, and it will sink down amongst them in its last rites like a crumbling megalith.

I'm seeing it through a mist now, astonished that I could be so moved by a vegetable. I back away a little, and look at it through my binoculars. I'm trying to frame it as a picture, the old Picturesque discipline. But the bony pterodactyl in its halo of green does not look like any ancient landscape painting. It's defiantly modernist. It could be a Miró squiggle, or a bizarre surrealist coupling, or abstract expressionism gone three-dimensional. It could be one of Maurice

Cockrill's doors, with the new green forms emerging from the shadowed pit. But mostly it makes me think of the paintings van Gogh made in the last months of his life, those total immersions in the chaotic creativity of nature which make no concessions to our tidy-minded perceptions. It insists that natural systems are never *completed*, not contained within fixed time-frames. Uvedale Price was right to connect the grammar of painting with the grammar of nature, but not to suggest that we need the first to appreciate the second. What 01243 says is primitive, sympathetic, universally recognisable. It's both reality and metaphor, a living instance of nature's resilience, and of the graciouness of survival. Simon Schama, in his uncompromising TV series on the *Power of Art*, said that art is about learning what it means to be human. In the more inclusive arena where we're now trying to live, art – and the natural forms that spontaneously aspire to it – could also be said to be about learning what it means to simply be alive.

III

The man who was chiefly responsible for reviving interest in the Picturesque was Christopher Hussey. He lived in the grounds of Scotney Castle in Kent in the 1930s, in a house built by his grandfather, Edward Hussey, from stone recycled from the castle. Edward, a devotee of Price, had abandoned the old house because it was damp, and then deliberately ruined it to make it into a 'picturesque object'. The garden had been 'improved' in 1837, and it was to be exactly 150 years later, when Christopher was long gone, before it too was rendered into a picturesque object one famously blowy mid-October night.

His distinguished study, *The Picturesque: Studies in a Point of View* (1923), contains many insights but none more striking or relevant to understanding our perception of trees than when he says: 'the picturesque interregnum between classic and romantic art was necessary in order for the imagination to form the habit of feeling through the eyes'. Feeling through the eyes! It's the kind of phrase that stuns you, makes you reframe old assumptions. Yet human eyes were able to receive affective signals from the landscape long before they recognised them in paintings. That is what *made* humans paint, what first generated those electrifying images of nature on the cave walls of southern Europe more than 30,000 years ago.

The following summer, on 20 July 2006, I was back in the Chilterns. It was the morning after the hottest July day on record, maybe the hottest since the Ice Age. The temperature had scarcely dropped a degree from yesterday's high of 36°C. Everything seemed glazed, embedded in hot amber. I walked about in slow motion, feeling my senses doing the same. Sounds were muffled. My eyes were dehydrated and prickly. The baking air had banished all delicacies of sensation. Any smells beyond dust and parched grass had evaporated by ten o'clock. This was sensory deprivation, courtesy of global warming.

I'd gone back for a purpose. I was still looking for resonances between biology and beauty, and wanted to see if they went beyond the 'theatre of the eye', and down into the invisible depths of the beech tree's underground existence. Specifically I was on the hunt for a ghost orchid, *Epipogium aphyllum,* Britain's rarest and most mysterious plant, and a flower whose capriciousness seems part of the spirit of the beechwoods. It's ethereal and shape-shifting. It grows rooted in deep beech debris in the shadiest parts of the woods, and is called the ghost orchid because it vanishes for decades at a time.

The heat was making me wonder if my plan was a bad mistake. I didn't know what it might feel like to be in a deep beechwood at 35°, starved of air, disorientated by those endless verticals, plagued by horse-flies. But I needn't have worried. When I arrived at the wood down in the southern Chilterns, it must have been at least 5° cooler in the shade, and the clarity of the beech trunks brought my eyes back into focus. The space was elemental: brown floor, grey uprights, deep green ceiling. No yellowing of the leaves, no sign again of the condition of beech deteriorating. The only sound I could hear was the crunch of my own footsteps in the dried-out leaf-litter. Hugh Johnson reckons that 'an old beechwood has the longest echo of any woodland . . . the echo of an empty room'. That was pretty much the view I'd held in my thirties, the snobbish opinion that high forest beechwoods were an ecological desert. I know better now, and understand that their liveliness is a cryptic quality, hidden, so to speak, in different dimensions. One of these is under the ground.

I'd picked my wood carefully. It was one of the sites where the ghost orchid had manifested itself during the previous decade, most recently in 1998. No one is sure why it has such fugitive habits. It has no chlorophyll and depends for food on a network of underground fungi. They're impacted with the orchid's roots, and break down nutrients from the mat of rotting leaves and wood that covers the floor. The orchid spends most of its life entirely below ground, coming erratically into bloom every ten or twenty years. Sometimes, quite indifferent to the universe of light, it blooms *under* the soil. When it does appear, the flower is phantasmagoric, like a pink sea-creature, a prawn dangled on the curling barnacle of the stem. Michael Longley's poem 'The Ghost Orchid' glimpses a vegetable so mysterious and sensitive that it can barely have an existence outside the imagination:

> Added to its few remaining sites will be the stanza
> I compose about leaves like flakes of skin, a colour

Dithering between pink and yellow, and then the root
That grows like coral among shadows and leaf-litter.
Just touching the petals bruises them into darkness.

I'm jealous of Longley if he'd glimpsed the plant himself. Only about 50 flowering spikes have ever been seen in Britain since the first record in 1854. In a strange tradition, common to several rare orchids, all the early sightings were by women. Mrs Anderton Smith found it in a small oakwood (its more frequent habitat on the continent) on the banks of the Sapey Brook in Herefordshire. She removed it to the safety of her garden where it promptly died. Twenty-two years later, a Miss Lloyd 'found a specimen' further up the Welsh Marches, growing in Bringewood Chase near Ludlow. 'Miss Peele' found the plant again in 1878, growing in the same wood, and then again in 1892, the specimens being verified by 'Miss Lewis, of Ludlow'.

It appeared once more at another Herefordshire site in 1910, and then vanished from this part of England altogether. It re-emerged in 1924 in a Chiltern beechwood west of Henley, where it was found by a local schoolgirl, Vera Paul, and confirmed by 'Miss Holly' and the distinguished botanist Claridge Druce. Seven years later it was discovered in a wood in nearby Satwell, growing out of a decaying beech stump. It wasn't found there again until 1953, but then reappeared spasmodically – and in another beechwood nearby – until 1999.

How could such a species survive, scattered across the country, and across the centuries? When I arrived at the wood, I had no real expectations of finding one. I didn't even know how one set about looking for such an elusive life-form. I drifted about, crouching down every few yards to scan the ground with my binoculars for any splashes of colour or vertical lines that broke the horizontal brown haze. A woman, walking her dog, asked whether I was bird-watching. I explained about the ghost, but she, not in the lineage of orchid dames, had never heard of it. I wasn't any luckier. I saw false ghosts and orchid spooks aplenty. The pale wisps of dried-out bluebell stalks, yellowing shoots of dog's mercury, once a scrap of lurid plastic that was the very image of an *Epipogium* flower. But not the real thing.

I tried widening my hunt, and began peering for other beechwood orchids. When I first came to these southern woods thirty years ago, eyes firmly down then as well, I used to find them every visit. Common helleborine, violet helleborine, narrow-lipped helleborine – *Epipactis leptochila,* an exquisite minimalist orchid with flowers the same pale green as young beech leaves. But there was nothing. Was I losing the diviner's touch? Looking for plants in the wrong place? Or was this another dimension of the wood that had been burned away by the long heatwave? The orchids' dependence on their fungal partners means that they're exceptionally sensitive to the drying-out of the soil, a condition in which fungal metabolism slows down. In fact the entire woodland relies on the immense network of fungal 'roots' (mycelia) which lies beneath it. The toadstools we see on the surface are just the fruiting bodies of a subterranean forest of tangled feeding threads, living in symbiotic relationships with the trees and many other plants. Some penetrate deep into the cell structure of the roots. Others form a kind of sheath round them. The fungi extract sugars which the trees manufacture by photosynthesis. In turn they help break down the rotting leaves and wood on the forest floor, and make the nutrients available to the trees and other ground plants. A tree like the beech can't survive without its fungal partner, and one of the main reasons for the sudden death of seedlings is that they have failed to find one in time. A single underground fungus may cover many hectares (and weigh several tons), and link all the trees in a wood, regardless of their species. The whole system is a feeding co-operative. I can't imagine the aesthetes of the Picturesque would have thought much of the ghost orchid's looks. But they might have been awestruck by the notion of this submerged fungal forest, mirroring the intricacy and variety of the twig-work above.

Leaving the wood in the late afternoon, I let my eyes stray up for the first time, to the great whalebone arches made by the high branches, laced together 30 metres above me. They were reputedly one of the inspirations behind 'arboreal Gothic' architecture. But for me, just for that moment, they looked more like the bottom of an upturned boat.

In the end the long lianas of the Picturesque, and probably a few Gothic tendrils too, reached Hardings Wood. I decided it was time we did something with the beech plantation, and began devising a plan to thin it. It had been planted for timber after all, and we would simply be fulfilling its destiny. A bit more light might encourage some regeneration, and maybe incursions by the flowers and ferns of the old wood. I used the high-minded phrase 'continuity of intention' to make the scheme respectable to myself. But muddled up with it were other kinds of intention that all too clearly echoed the projects of the 18th-century woodland philosophers. I nodded towards the Earl of Caernarvon's shameless belief that woods were there to provide for the payment of debts, and resolved to sell the thinnings once I'd cut them down. I admired Uvedale Price's vision of woods 'full of openings and hollows; of trees advancing before, or retiring behind each other – all productive of intricacy, of variety, of deep shadows, and brilliant lights', and thought that I could make that happen in the plantation – with enough heavy machinery.

This wasn't the kind of project we could do as amateurs. We needed expert help. So I found a sympathetic local estate agent to mark up the wood for felling. Ken and I went up there together one November afternoon. The basic principle, he explained, was to mark the trees you wanted to keep first: 'What you get depends on what you take.' He would choose the ones that might make decent timber trees, and I could choose those that were picturesque, or promising dead wood habitats, or had just tickled my sentimental fancy. So we progressed round the plantation, in a strange parody of John Toovey and his girl assistant's procession round the woods at Shirburn, squirting puffs of yellow aerosol paint at a bizarre combination of majestic pillars and forked eccentrics. Then we marked what foresters call rubbish – trees which weren't up to either of our purposes. Dozens of boringly crooked and under-achieving trees were thus summarily condemned, and robbed of their chances to make astounding responses to the next storm, or turn into another number 01243. I consoled myself by thinking that these were the trees that would probably have died from shading out anyway. I was pleased to learn that this fastidious picking-out of individual beeches to bring about a more promising mix of ages was known as 'selection'. It sounded as if we were preparing a Tree Show for the Royal Academy.

Then came the difficult part, marking up the trees (in red paint this time) that might comprise an instant money-making harvest, but doing this mostly around the peripheries of the chosen ones, to give them more space to flourish. One group were uniform enough for us to mark out a potential clearing of about a quarter of a hectare, where we might get some real regeneration. And, of course, I always had in my mind the provision of some picturesque variety on the way, so that, in Price's words, 'in walking about . . . the form changes at every step; new combinations, new lights and shades, new inlets present themselves at every step'. I found I was rather relishing this wide-screen enlargement of my personal tinkerings in the old wood, and the chance to bring together a personal fantasy, a wood on fast-forward, a stab at ecological enhancement, and a small down-payment on a new pick-up truck, all, so to speak, in one fell swoop.

So we agreed on the fates of about 100 beech trees, and Ken set about the task of finding someone to cut them down and sell them. This didn't prove easy. Most contractors weren't interested in small jobs on difficult terrain, and the market for beech thinnings was still glutted after the 1987 storm. But Bob, a local freelance with a great knowledge of the Chiltern beechwoods and original thoughts about how to winkle single trees out of a dense throng, agreed to take the job.

He arrived late in the winter, with nothing more than an old Ferguson tractor, and set about the task entirely by himself. He felled most of the trees without ropes, slipping them between standing trees just by skilled use of the chainsaw. He took the tops and side-branches off and dragged the trunks one by one behind his tractor, out of the wood and into a field by the lane. He worked on into the dark, and sometimes I'd spot his headlights at midnight, weaving through the dark shadow of the wood like will-'o'-the-wisps. I worried about him, especially after he told me of a colleague who had pollarded himself when he slipped from a branch.

But the job was finished without a scratch either to Bob or to any of the remaining beeches. The felled trunks, each one chalked with its own personal number, lay in the field below the wood, ranged side by side like a flotilla, and waiting to be collected by a timber merchant and ferried to the furniture factories in High Wycombe. I used to gaze down at them from inside the wood, thinking how much they resembled the rafts of trunks slid down into rivers by real lumberjacks. The plantation looked transformed, ripened into natural-looking maturity in a couple of weeks. The change in the light was astonishing. The wood seemed immense, impossibly high, dusted with gold as sunbeams tracked through the new gaps in the canopy.

What remained for us fledgling village lumberjacks was to make good. We had to become woodland vultures, and clear up the offal beech. And we loved it. Logging up beech branch-wood was a cut above dealing with sappy ash and sodden poplar. The wood was as crisp as the crust of a new loaf, and smelt like it too. The cut logs clattered onto the growing wood-piles with the ripple of a xylophone. These stacks became the objects of huge pride, as if we were trying to echo the tidy new geometry of the plantation – or make amends for it – by putting its lopped bits back together. The art critic Yvette Wiener, in her introduction to a collection of remarkable photographs of the traditional woodstacks of Austria's Ausseerland, tries to understand this paradoxical habit:

> What is the inner image which informs each stacker, what internal pattern is he attempting to reproduce in his careful elaboration? What makes one assemble the same regular round shapes whilst another puts together only square, beam-like pieces and yet another mixes them carefully, framing one with the other? Who is the mad man who meticulously reassembles in the stacking the log he has split beforehand, giving it back its original shape?

Feeling through the eyes. Like the polled beeches of Burnham, we seemed impelled to try and hold the centre, to strive for rightness in a tree, even when it is split into pieces and lying horizontal on the ground.

Chapter Six: The Natural Aspect

The 19th century. Landowners are raiding the last redoubts of the wooded commons. The trees fight back, via eccentric cohorts of liberal lords and roughneck commoners in England, and bohemian artists in France. Out of the mêlée come new images of 'the natural'.

I

EACH TIME I visit Frithsden Beeches I follow the same path, winding my way between the edge of the grove and the open common. Grazing stopped in the 1920s, and the heath is now well on the way to becoming a young oakwood. If I wander to the right I'm crunching last year's beech leaves underfoot. A few yards to the left and I'm treading on dead bracken fronds, a softer noise, like the folding of linen. Two different places still separated by a whisper. There are these soft sounds, but not much colour. It's a flat afternoon in March, too early for the full blaze of gorse or the glow of the first beech leaves. Under the trees I can just make out a group of fallow deer, descendants of the herd that was introduced by the Normans in the 11th century. With no leaf-shade for cover, they're warier than usual. They stand stock-still, staring at me, their soft fawn bodies seeming like buttresses against the curves of the beeches. By chance they're hiding out in what may have once been an animal shelter. A shallow bank traces out a circle about 80 metres wide, joining up a scatter of low-slung pollards, multi-fluted columns. The main branches all project outwards from the circle, like elbows. The inside growth must have been browsed out centuries ago. I wonder whether there was a fence on top of the bank once, against common law and custom? The perimeter of a pen for the deer maybe, a lord of the manor's illicit perk?

There have always been shenanigans up here. Ambitious chancers trying their luck, big landowners extending their domains. In an earlier March, in 1866, the Beeches and much of the land that surrounds them were the site of an historic battle against enclosure. The events had all the flavour of a Victorian melodrama: a sickly young lord of the manor and his autocratic mother, an audacious night-time raid by a gang of hired mercenaries, and a sensational court case that helped shape the future of England's commonland. It was a critical moment in the fortunes of ancient woods. Up till then wooded commons throughout southern England had suffered a long history of attrition and illegal enclosure, which reached a climax during the 'improvements' and plantation fevers of the late 18th and 19th centuries. Historically they almost all lay on poor soils, of little attraction to farmers, though they were hugely productive in non-agricultural ways. The great common of Berkhamsted, for instance, which contained the Beeches, was shared between eleven surrounding parishes. Their inhabitants grazed cattle on the heather, put out pigs in autumn for beechmast and acorns, cut gorse for bread-ovens, beech-wood for their fires, and bracken for cattle bedding. Not profitable livelihoods, but independent ones, lived out necessarily with mutual respect for each other and for the landscape.

But during the 18th and 19th centuries the human population was growing inexorably. There were increasing pressures on land, for food and housing. New techniques for manuring soils and growing trees *en masse* were developed to push up production. Easy access to coal was making firewood redundant, and wooded commons began to look, in the improver's eyes, doubly irrelevant. They were not just wasteland, but wasted land. It was as if their grotesque pollards – fey, superannuated, inefficient trees, symbolising the decrepitude of economy and morals that persisted out of the reach of absolute property rights – were asking for it.

Yet in response to these economic and social developments, this was also the period of a burgeoning new affection for picturesque nature and the open air, for green lungs for the city dwellers. Ancient trees figured strongly in this vision of new arcadian retreats beyond the smoke. So two centuries of argument about the respective merits of beauty, utility and naturalness, and about the ancestral rights of commoners and landowners, were on a collision course.

In 1865, the Commons Preservation Society was formed, a body which, in the words of one of its founders, Edward North Buxton, was 'fully impressed with the importance of providing, for those who live the artificial life of our great city, the means of studying nature where it is unrestrained by art' – a noble aim which contained within itself the contradictions it was hoping to resolve. The following year matters came to a head simultaneously in commons across the south of England. Berkhamsted Common and Epping Forest were enclosed, and then liberated. The New Forest Association was formed to oppose the enclosure of the ancient woods and heaths (leading to an Act eleven years later which, amongst other objectives, required the conservation of 'the picturesque character of the ground'). And at the end of a climactic year, the Metropolitan Commons Act was passed, which protected all commonland within 20 miles of large urban areas, and guaranteed the general public the right of 'air and exercise' over it. That the beech tree was a central character in this drama was partly coincidental: it simply happened to be one of the most frequent trees on land not much good for anything but commoning. But there was something about its outlaw character, its dogged but inventive adaptability, that made it the natural mascot in a battle on behalf of wild land and ordinary people.

II

The Normans had their eyes on Frithsden Beeches from the start. Soon after the Conquest, William had set up base in Berkhamsted's ancient castle to mark time till his coronation. It was a comforting retreat from which to contemplate a new kingdom, and surrounded, like no other castle in England, by three moats. From the top of the keep, he could see the beech savannah that stretched out to the north. A tempting prospect. Vast lands for hunting, for timber, for generating taxes. Twenty years later, in the Domesday Book, the Normans valued it as 'wood for a thousand hogs', and moved in their own pigs. Later still they annexed part of it as a deer park.

In 1300 an inquisition post-mortem gave exact details of this wooded part of the Normans' new estate, and the rights they had granted over it: 'A certain wood called Del Frith, which contains in itself 763 acres [308 hectares] and one rood [about 6 metres]; and a common as well for the freemen as the villeins of Berkhamsted excepting pannage time namely between Michaelmas and Martinmas.' Pannage was the pasturing of pigs in woodland, during the period when beechmast and acorns were on the ground; the Normans had appropriated for themselves, and for the monastery at nearby

Ashridge, what was one of the Frith's most valuable asset. The ordinary townspeople and commoners of Berkhamsted and the surrounding villages had to pay pannage dues for their animals. And in 1353, Edward the Black Prince (then Earl of Cornwall and resident in the castle) decided to make the *cordon sanitaire* more secure. He made plans to extend his park, and enclose the whole area 'with a paling for the preservation thereof and of the game there'. He needed weather-proof oak for this fencing, of which there was only a smattering in the Frith. So the prolific beeches – useless as outdoor fence-posts but peerless as fuel – would have to become currency, to pay for the job. The prince's steward was instructed to

> sell beeches in the foreign wood of Berkhamstede to the value of £20 and employ the money in the purchase of oaken timber for the rails and stanchions of the said paling, cutting down suitable beeches for the remainder thereof. All the beeches are to be cut down in different parts of the wood, as shall be most profitable for the prince and least wasteful of the wood, and the stumps are to be marked with the axe appointed for the purpose.

The Norman description of the Frith as 'the foreign wood' was uncontentious (it simply means the land outside the manorial park), but with hindsight seems to suggest the contempt the Normans held for the local people, and their desire to keep them at arm's length.

For the next decade the prince, always extravagant in his habits, continued raiding the Frith. In 1358 he sold 200 beeches, and the following February another 100 'in order to make the final payment for the works in the castle'. Three years later he marked – banked, in effect – 600 more for his exclusive use. His charitable gestures were also made in this local and seemingly inexhaustible resource. He gives six trees to the warden of the hospital of St Thomas, four to the Friars

Preachers of Dunstable, one to a 'poor man at Aylesbury called Peek'. Beeches are given as a Christmas present to Berkhamsted's vicar; as a reward for a loyal service to Simon de Driffield, yeoman of the scullery; to the rector of Ashridge 'in return for some fuel which was used in the burying of the Prince's clerk, Sir John de Hale', and to Thomas Bolitter 'in recompense for a boat which he lent to the prince and which was broken in the prince's service'.

What kind of gift were these trees? Did the prince's men simply hack them down and cart the logs over to the lucky beneficiaries? Or were the gifts more subtle, an early gesture towards woodland conservation? In parts of the continent where there were similar wood-pasture systems, named individuals sometimes had the lopping rights to particular pollards, or, in the case of especially big trees, to just one side of them. Might the prince have donated beeches which were still standing? Potential pollards, do-it-yourself fuel lots? Whichever, he was a risk-taking businessman. In 1362, a violent storm flattened many of the beeches that remained. The prince immediately ordered them to be sold as profitably as possible, neglecting to remember that fallen trees and branches were the prerogative of his parker. John de Newenton was bought off with a gift of 100 shillings for the loss of his perk.

This rapid consumption of beeches was not so much selling off the family's silver as poaching one of the commoners' staples. The Frith, always open to grazing, shrank and turned to heathland across much of its extent, the beeches surviving as pockets of old trees. And a smouldering resentment against appropriation settled in the folk memory of the local commoners.

History like this, of course, just happens. There is no plot. The Norman line that became the Duchy of Cornwall were doubtless just trying to cope with the exigencies of the cost of living, keep their employees quiet, and grab as much land for themselves as possible. But seen not so much with hindsight but, so to speak, site-sight, from inside the Frith, the story looks more ominous. It's a narrative of liberties taken, in which the Beeches seem persistently cast as victims. They're repeatedly looted, excoriated as bad influences and wastes of land, granted reprieves, turned into scapegoats and hostages. The evidence can be cut many ways, depending on your sympathies. This is the story of the Beeches with an eye for the trees' experience of what happened.

The year 1607, a time when the Queen Beech was quite likely a sapling. Neither the Crown, nor some pilfering commoner, but the agent of the Duchy of Cornwall (now the landlord) was caught poaching 120 beeches in the Frith. The Duchy, hoping to save face and entrench its own position in one swoop, set up an independent commission of inquiry into the status of the common, a jury of tenants headed by Sir John Dodderidge. But the findings didn't swing their way. The jury concluded that the Frith was ancient commonland, never enclosed 'within the memory of man', and that all the inhabitants of Berkhamsted (not just the manor's tenants) had rights of grazing and pannage, and the gathering of fern, furze and firewood. The Duchy ignored the judgment, and two years later proposed to enclose part of the common for its own benefit, probably with the intention of turning the heathland into arable, and converting the pollard groves into timber lots. Years of intimidation, lobbying, backroom deals and bribery followed. In the end a rough kind of agreement was reached to enclose 300 acres of the open common, but to preserve the beech-wood parts. The new farmland estate was named Coldharbour.

That, the Duchy solemnly promised, was the limit of its territorial ambitions in Berkhamsted. But just twenty years later, it was at it again, on behalf of the future Charles II (then the 9-year-old Prince of Wales), making further bids 'for the improving of the waste and commonable grounds'. This time the Duchy engineered the support of the Church and a minority of local landowners in a scheme to enclose a half of what remained of the common (about 400 acres, including the Beeches) leaving 100 acres 'for the use of the Poor', and went as far as putting in live hedges, palings and ditches. But in August 1640, a local man called John Edlyn led a revolt of the local commoners against the enclosures. In his own words, they 'were interupted by the generallity of the Inhabitants and Tenants being at least six thousand, by some of which the enclosures were throwne downe and lay'd open as before'. The work was done at night, in the presence of a huge crowd of onlookers, by a disciplined force of a hundred men under Edlyn's supervision.

In 1761, the Duchy leased all the profits of the common to the Earl of Bridgewater, for services to his country. The service in question was the completion of the first navigable canal in Britain – one of whose incidental consequences was to bring cheap coal to the area, and end the long reign of beech-wood as the chief local fuel. The Grand Union Canal reached Berkhamsted in 1799, the railway forty years later, and pollarding of the Beeches, to judge from ring-counts I've made on fallen branches, ceased in about 1850.

The Bridgewaters had owned Ashridge Park since 1604, which gave them rights over the adjacent commons of Berkhamsted. Like the Duchy in those days, they were an expansive and unscrupulous bunch. For the best part of a hundred years successive earls had been surreptitiously enclosing small patches of woodland round the perimeter of the common. They'd fell the profitable timber, fence the open coppices, or, if they were too big, engage in a little backstage politicking with the assistance of journeymen lawyers. No attempts were made to consult the main body of commoners.

Near the end of his life, when he'd become fabulously rich as a result of the expansion of his canal network, the duke decided to rebuild the decrepit Ashridge House as a Gothic mansion, and monopolised the gorse on the common to feed his brick-kilns. He died in 1803, and his son, the seventh earl, completed the project. He then developed a passion for road-building, stopping up two ancient trackways across the common, and building new ones for his own convenience – and for his wife's, who was fond of horse-riding in scenic and comfortable surroundings. Her friend, Lady Marian Alford, shortly to become a key player in the unfolding drama, wrote:

> When we first visited Lady Bridgewater, shortly after our marriage in 1841, she took us to a ball at Berkhamsted, and we were guarded by outriders with loaded pistols in their holsters . . . As the woods and forests of Italy were usually the refuge for brigands and outlaws, so in a greatly modified degree the unenclosed Commons had become the resort of the unfortunate and least respectable members of the community. Few took their pleasure there except on horseback.

Not perhaps the best viewpoint from which to observe the local people who made a living there.

The long reign of the Duchy and the Bridgewaters began to peter out in 1823 with the death of the third earl. He left his estate to Lord Alford, husband of the petulant Lady Marian, whose young son, after complex legal wrangling, became Lord Brownlow in 1853. He was 11 years old, and far from well, and his mother remained his guardian and, in effect, the manager of the estate. She kept a journal, which provides an insight into the patrician view of natural landscape. In 1854 she reported that a deputation from the town had visited her, expressing their wish to have the beechwoods and commons grubbed up for cultivation.

> The memorial laid before me represented that such a measure would be for the benefit of the poor of the town of Berkhampstead [sic] and of the Ashridge Estates. I replied that our rights over these commons were more than shared with the Duchy of Cornwall, and that I could not, during my son's minority, consent as a guardian to the disfigurement of the beautiful Frith, which I considered as the glory of that part of the county. I reminded the deputation that Linnaeus, when visiting England, came to Berkhampstead, and seeing the gorse in full blossom knelt down and thanked God for showing him so glorious a sight. From that time I heard no more of spade-labour from the people of Berkhampstead.

This from the woman who thought the common was only pleasing from the top of a horse. It's unclear who made up this deputation, but it's inconceivable that, as Lady Marian later insisted, they represented the people of Berkhamsted, most of whom had no formal rights over the common and would get no benefit whatsoever from it being turned over to agriculture. But in 1863 she effected, on her son's behalf, the purchase from the Duchy of the entire Manor of Berkhamsted, including the common and all rights over it, for £144,546, an immense sum of money then. Almost immediately, the new lord, who had just come of age, set about the place, building a new road, digging trenches across the old trackways, fencing plots of land alongside. He'd obtained the consent of the parish vestry for the road, but they were under the impression that it would simply be an addition to the common, not a device for shutting out the public. Brownlow's agent, William Paxton, then began buying out the rights of as many official commoners as he could (these were smallholders and tenants with property adjoining the common, or whose holdings had inherited rights attached to them). In return for surrendering their customary rights over many hundreds of acres, Brownlow offered the town, as 'liberal compensation in lieu of [their] trivial outstanding claims', a few acres of swamp for a recreation ground down by the railway line . . .

The town was in a moider, and feared the worst. But one large local landowner with common rights had stood firm against Paxton's blandishments. Augustus Smith, owner of Ashlyns Hall, Lord of the Scilly Isles, and widely regarded as a progressive and honourable gentleman, had received an offer for his common rights from Brownlow's solicitors on 25 October 1865. He'd made a generous reply, saying that though he greatly regretted the enclosure of commons in the neighbourhood, 'material improvements cannot well be contended against'. But he was unhappy about the underhand methods Brownlow was employing, and suggested that the whole affair be put in the hand of the Enclosure Commissioners in the usual way. No talk of politics or people's rights, just propriety. On 10 November, Brownlow agreed to put his plans before the Commissioners. But on 3 January, writing from his convalescence bed in Menton in the south of France, he said that this was not what he had meant at all. Then, spurred on by the impatience of his ambitious mother, or perhaps the suggestion of his agent (William Paxton was the brother of Joseph, creator of the Crystal Palace, and maybe had a desire to create a monument of his own), he took matters into his own hands. In February, Paxton supervised the erection of more than 3 miles of iron fencing round 400 acres of the common. The barricade was 6 feet high, with seven horizontal rails, and contained not a single opening or gate. It enclosed the whole of the Beeches, several other groves of ancient pollards, and an area of heath to the east. Lady Marian explained later that their intentions had been to enclose and cultivate all this eastern area, and plant up part of the wood-pasture 'for shelter and ornament' – a scheme which would have doubtless spelt the end of the Beeches. She protested that it had all been done because of the young Earl's 'desire to make every experiment towards benefiting the poor'; and when asked why they hadn't announced their intention in advance, simply quoted from one of the family's mottoes: *Esse non videri*, 'To be not to seem'.

This high-handed and almost certainly illegal appropriation of the common caused outrage in the district. It was obvious that Lady Marian and the Brownlow team were being disingenuous, and that their pious concern for the welfare of the common and the local poor was no more than a cover for a self-interested land-snatch. It also seemed that the family had no regard for the 'trivial outstanding claims' of the townspeople, who, as the inquiry of 1607 had ruled, had their rights to cut fern and gorse simply by virtue of being inhabitants, not property-owning commoners. But mid-19th-century country people had little of the political verve of their 17th-century ancestors, and weren't willing to take on a powerful landowning family like the Brownlows.

At this point the recently formed Commons Preservation Society was asked for its advice. Fresh from their victorious rescue of Hampstead Heath, its officers reckoned that what was needed was the co-operation of a commoner with a 'long purse, and with independence and courage' to challenge Brownlow in the courts. Augustus Smith, already embroiled in the affair, was the obvious choice. Lord Eversley, the founder of the society, and its solicitor P. H. Lawrence, met Smith at the House of Commons, and in the course of their discussion decided to side-step the law in the same way Lord Brownlow had, and 'to resort to the old practice [highly successful in 1607] of abating the inclosure by the forcible removal of all the fences, in a manner which would be a demonstration, and an assertion of right, not less conspicuous than their erection'.

It takes one's modern political breath away: two lords of the realm, seated in the mother of Parliaments, are conspiring to take violent action against the property of a third, all on behalf of the timorous citizens of a small Chiltern town, and their gorse bushes and bracken-lapped beeches. No wonder that what followed had all the character of an outrageous – but exquisitely rousing – political comedy. Smith's next move was to hire a large gang of physically deft wreckers. At midnight on the evening of 6 March 1866, 120 navvies recruited by his London contractor assembled at Euston Station, for a train-ride north-west. Their crowbars and sledgehammers had an entire carriage to themselves. The train reached Tring, the nearest station to the common, at 1.30 the next morning. There the comedy continued. The contractor was missing. He'd got senselessly drunk on his fee at Euston Station and had not made it to the train. Fortunately the society's lawyer, Mr Lawrence, had sent a 'confidential clerk' with the party, and he took charge. So the procession of road-workers, railwaymen and chippies began the 3-mile march to the common, doubtless feeling some pangs of envy for their erstwhile boss as they passed, long after closing time, the Royal Hotel. (There was some poetic justice in this small geographical detail. The Royal was where a clique of local landowners had schemed over the enclosure of the adjacent parish of Wigginton, including Hardings Wood, just thirteen years before.)

Lawrence's man explained the purpose of the expedition to his band of guerrillas as they walked up the hill towards Aldbury, with its own pollard beeches shining silver in the light of a full moon. When they reached the edge of the common, they split up into smaller gangs and began dismantling the fences. By 6 o'clock the railings and posts were down, laid out, as an eye-witness newspaper report put it, with 'each stout upright having the metal bands first neatly folded round it, and then being laid on the turf it had recently served to close in', a careful avoidance of any hint of gratuitous damage. 'Meanwhile,' the report continued, 'the news spread, and the inhabitants of the adjacent village and district flocked upon the scene. In carriages, gigs, dogcarts and on foot, gentry, shopkeepers, husbandmen, women and children, at once tested the reality of what they saw by strolling over and squatting on the Common and taking away morsels of gorse, to prove, as they said, the place was their own again.' Brownlow's representative, Walter Hazell (a grocer regarded locally as a turncoat), didn't arrive on the scene till 7 a.m., but was 'too late to do more than protest against the alleged trespass, and this was energetically done'. But *Punch*, which a fortnight later published a riotous sixteen-stanza ballad, very much taking the townspeople's side, was less discreet:

'Hurrah!' the navvies shouted:
In sight a horseman glides:
See on his cob, with bob, bob, bob,
The duteous Hazell rides:
To do his Lordship service
Comes riding through the mirk
And bids the navvies let him know
Who brought them to their work.

Answer the stalwart navvies,
Who smoke the ham-smoker's game,
'Behold'st though, Hazell, yon canal;
Would'st like to swim the same?
If not with beer this instant
Thyself and cob redeem,'
And round him as they spoke, they drew
And edged him near the stream.

So down went Brownlow's railings,
And down went Hazell's beer,

And from the gathering crowd upgoes
One loud and lusty cheer . . .

The very next day Lord Brownlow issued a writ against Augustus Smith for trespass and criminal damage, but then tried to persuade him to settle out of court. He held back, increasingly uncertain about the strength of his case, and then died the following February, aged 25. The action for trespass was never heard. Meanwhile Smith had issued a counter-suit, asking for an injunction to prevent the Brownlows from enclosing any part of the common in the future, and for a court ruling on what rights existed over it. The action dragged on for nearly four years, and involved oral evidence from local people and tangled legal reviews of documents going back to the early 17th century. Smith was asking for a whole range of rights to be authorised, including grazing, cutting fern and gorse, and 'estovers and haybote and housebote' (the rights to lop beeches and other trees for fuelwood, and for the repair of houses and fences). The suit also claimed 'a right to use the whole of the Common . . . for the enjoyment of air and exercise, and for amusement and recreation'. But, oddly, no evidence was offered on these last two points, and when Lord Romilly, Master of the Rolls, gave his judgment overwhelmingly in favour of Augustus Smith and the commoners, he specifically omitted rights of estovers and recreation. So though the whole affair had been a triumphant victory for Smith, and 'one of the most decided and vigorous protests against . . . usurpation which have occurred in our prosaic, peaceful and order-loving times', it fell short of what the Commons Preservation Society was seeking. *The Times* agreed. 'Lord Brownlow probably does not expect popular sympathy to be with him in the defeat of his attempt to enclose Berkhamsted Common' their leader pronounced. 'He certainly will find none, [but] a comparison of the ostensible objects of this battle in the Court of Chancery and its results shows one formidable item in the list of killed and missing. We should have been sorry for the pigs of Berkhamsted to have had any of their rights over the acorns of Berkhamsted Frith curtailed. The public interest, however, centred . . . in the claim which Mr Augustus Smith set up to have the entire Common treated as a recreation ground . . . The public may be able to prove the enjoyment of rights of way; but the necessity of asserting against Lords of the Manors its right to keep waste lands as breathing spaces for growing towns is too recent in date to have created precedents for resisting interference with their uses of this sort.'

Berkhamsted Common remained open and grazed until the 1920s, when it was sold off by the Brownlows. The local golf club bought 121 hectares, and the National Trust the remainder, including the portion which contained 'del Frith', and which had, against all probability, survived more or less intact during a thousand years of wrangling.

III

In the same month as the night-time raid on Berkhamsted Common, Samuel and Alfred Willingale, two labourers from Loughton, Essex, were each fined *2s 6d* for damage to the ancient beeches and hornbeams of Epping Forest. In a drama which echoed the Berkhamsted case, they had taken direct action to establish their ancient rights of lopping, in response to the fencing of 532 hectares of the Forest by the lord of the manor of Loughton the year before.

But the case of Epping was different from Berkhamsted in many ways. Epping was huge – 2,428 surviving hectares of Royal Forest – and a common with a complex system of ownership and rights. One of the most fiercely defended of these had always been the right to lop the hornbeams, beeches and oak for firewood, a practice which had died out in Berkhamsted. And Epping's closeness to London meant that it had been a public amenity – and a place of retreat – long before the Metropolitan Commons Act. In his fictional *Journal of the Plague Year* (published in 1722, but based in 1665) Daniel Defoe gives a vivid account of Londoners taking refuge in the Forest, believing that the greenery would in some way protect them from contagion, and prefiguring the much larger evacuations of the Second World War, when people fled there to escape the air-raids. Defoe's characters enter a forest rife with rumour and suspicion, and set up camp under a clump of pollards north of High Beech village, persuading the nervous locals to bring them food and furniture. Later in the 17th century the Forest became the haunt of ex-Civil War soldiers turned deer-stealers, and in the calmer years of the 18th and early 19th centuries, of dissenters and artists and poets. Tennyson lived at High Beech between 1837 and 1840, and spent a fortnight visiting (not as a patient) Dr Allen's local asylum, where he encountered John Clare (who was) composing his own anthem to the ancient *maquis*, 'London *versus* Epping Forest':

> The brakes like young stag's horns come up in spring
> And hide the rabbit holes and fox's den;
> They crowd about the forest everywhere,
> The ling and holly-bush and woods of beech . . .
> Thus London like a shrub among the hills
> Lies hid and lower than the bushes here.
> I could not bear to see the tearing plough
> Root up and steal the forest from the poor,
> But leave to freedom all she loves untamed,
> The forest walk enjoyed and loved by all.

Clare walked home to Helpstone in 1841. Dr Allen remained, finding his own peculiar inspiration in the Forest's trees. He'd invented a steam-powered machine for carving wood into home decorations – 'The Patent Method of Carving in Solid Wood' – and had persuaded Tennyson and his three sisters to invest in it. The scheme was a disaster, and both Allen and the Tennysons were bankrupted.

Yet if Clare had rightly read the development omens as city – or perhaps the City – versus forest, it was also true, as it had been at Frithsden Beeches, that the city's people were on the Forest's side. And if the two commons differed immensely in their size and status and history, their crises and salvations developed along almost identical paths. Both had been subject to a long history of attrition by the Crown and private landowners. Both aroused the moralisers' piety: 'the lopping of trees', one wrote in 1861, 'encourages habits of idleness and dislike of settled labour . . . which are injurious to the morals of the poor'. When the crisis came at Epping it was, as at Berkhamsted, precipitated by an illegal and opportunist act of enclosure by the lord of the manor of Loughton (who in this case also happened to be the rector of the parish). And again the Commons Preservation Society came to the rescue, backing the Willingale family's direct action, and searching for a powerful commoner who would be prepared to stand up in court on behalf of the Forest and its traditional rights. They found one in the Corporation of London, which owned a small area of land on the edge of the Forest, and for the next few years the Corporation fought their case in front of the Master of the Rolls. In July 1874, Sir George Jessel gave his judgment, and granted an injunction against the lords of the manors, prohibiting them from making any future enclosures, and requiring them to remove all fences erected within twenty years before the commencement of the suit – a judgment that returned 1,210 hectares to the open Forest. What he did not give, ironically, was any firm confirmation of the rights of all the commoners to lop the Forest's trees, the issue which had finally ignited the whole controversy. Neither the Corporation, nor the bulk of the witnesses to the Royal Commission on Epping Forest (which was progressing simultaneously), were interested in preserving that particular greenwood liberty. Their priorities were the conservation of the beauty of the forest, and its enjoyment by the people, and they regarded the 'periodical mutilation' of the trees as inimical to both ends. All these matters were to be dealt with in a forthcoming Act of Parliament.

As it happens the commoners' lopping rights were eventually proved, and they were paid a lump sum of £1,000 in compensation for imminent extinction of their ancient freedoms. A further £6,000 was awarded for the building of a community centre and reading room in Loughton village, to be called 'Lopper's Hall'. The very last cutting of Epping's pollards took place at the traditional time for the opening of the lopping season, midnight on 10 November 1879. Some 6,000 people gathered at Staples Hill in Loughton – including Lord Eversley, still fighting the commoners' corner. They perambulated the boundaries of the Manor by torchlight, and then went to lop the hornbeam and beeches till 2 a.m., when, according to custom, the logs were dragged out of the Forest on horse-drawn sledges.

The Epping Forest Act which outlawed this practice had been passed just a few months before. It put an end to the Crown's vestigial rights, and to all Forest Laws and 'burdensome customs'. The Corporation was to have the sole power 'to fell, cut, lop and manage in due course the timber and other trees, pollards and underwood', and yet the Act insisted that the overriding aim of these activities was 'at all times as far as possible [to] preserve the natural aspect of the Forest'. It had not foreseen that these means and ends might be incompatible, or that a 'natural aspect' – surely as plain and desirable as apple-pie – was in fact a goal as elusive and subjective as the Public

Good. What happened in Epping over the next years brought to a head two centuries of argument between foresters, scientists, artists and landscapers about the nature of beauty, about how woodlands regenerated themselves, and what that elusive word 'naturalness' actually meant. The arguments are still going on today.

The chief architect of the Bill and the major influence on how it was interpreted in action was Edward North Buxton, a member of a rich Quaker dynasty that had settled in the Forest in the 16th century. He set out his personal values in the 5th edition of his book *Epping Forest* (1898):

> It goes without saying that in a natural forest we should preserve those things which are not of man's doing. As an instance of this may be mentioned the importance of retaining trees which are decaying, trees which are dead, trees which have been overthrown by the forces of nature, as well as those which are in full vigour. I have recently spent a fortnight in exploring one of the largest natural forests in Eastern Europe. Here, to my mind, the chief beauty resides, not in the standing trees, but in the giants that lie prone among their roots. Many of them have lain for several centuries. They are gorgeous with moss and lichen; their great trunks are seed-beds for their descendants, and they tell a story of mighty hurricanes and snowstorms which we should miss, if it had been possible to remove them. Our Forest is also a document of nature with its tale to tell. Its ruins should be preserved, as well as its vigorous youth. It should not be trimmed and garnished.

This was a revolutionary vision, reviving Gilpin's and Uvedale Price's insights into the theatre of the eye, and the relation between beauty and history, and anticipating a libertarian attitude to conservation that few would dare sign up to again until the last years of the 20th century. And it was a faultless description of the state of naturalness in a forest. But that was not how things turned out in 'our Forest'. Epping *was* trimmed and garnished, chiefly because Buxton had not appreciated how nature produced the dramatic scenes he'd witnessed in eastern Europe.

He hoped that, given a kick-start, Epping would slowly progress towards something like an unmanaged continental beech forest. In this scenario pollards were, by definition, an unnatural intrusion, even though they most resembled the ancient 'ruins' that he admired. He wanted them out. As a philosophical position, this was impeccable, and in the very long term might have allowed the development of the 'natural' wood he longed for. The problem was that, for Buxton, it was overlain and compromised by personal aesthetic judgements. 'The mop-head growth', he wrote, 'substituted by the process of lopping, for the natural shape, is not only destructive of all variety and grandeur in the timber, but owing to the lodgement of moisture in the crown, and the consequent rotting of the heart of the timber, is fatal to the health and long life of the tree, and weakens its resistance to gales.' He was mistaken on all these last points of fact, but it is easy to sympathise with him about the 'mop-heads'. A black-and-white photograph of a stand of hornbeams near Staple Hill, taken just after they'd been lopped in 1877, is a pitiable scene, as shocking as the sight of a well-fleshed friend emaciated by illness. The trees, many of them quite young, have all been trimmed to the bone. They look like the blasted stumps in Paul Nash's terrifying First World War painting *The Menin Road.* One observer wrote in 1864 that this 'was no more nature's notion of a primeval woodland than are closely-cropped hair and shaven lip and chin her intentions for the real expression of the human face'. But the trees survived and grew back, as they have always done, and a century on have become the kind of elvish trees our

Picturesque-hungry spirits dote on. That's the paradox of pollards. We admire them when they're neglected – retired, so to speak. But it's only by being in work, being cut over, that they survive into the gale-resistant old age that Buxton so admired.

In 1893 he supervised the removal near Loughton of 'a considerable number of the ugliest of the pollard beeches, a measure which elicited a storm of criticism, but which nevertheless has, in my opinion, greatly improved the wood by giving more scope to the unpollarded beeches'. Again, a logically flawless move, even though it was bad news both for the Forest's aura of antiquity and for the organisms that had lived in these corrupted beeches for thousands of years. And it raised, perhaps for the first time, a conundrum which has become increasingly worrisome in our much-altered landscape: which is the more 'natural', to erase a human-shaped feature, or to allow it 'naturally' to evolve?

Buxton had no time for such subtleties. He knew the difference between 'very rugged and picturesque' stems and 'weaker and spindly sticks'. He had no doubt that 'the chief beauty of an oak lies in its massive lateral branches'; that, in general, the hornbeam was a 'meaner' tree than the beech. He was determined to let these ideal forms triumph. He insisted that he did not desire the removal of all pollards, many of which were 'curious and picturesque', but that, *en masse,* he found them 'monotonous and unpleasing'. He wanted, above all, to see the Forest full of young 'spears', as he forcefully described straight saplings, and which he knew the pollards were shading out. His epiphany in the dark forests of the East had manifested to him the energy of wild trees in the extremes of their existence. The majesty of the fallen giants, with their regenerative cloaks of epiphytes and seedlings, and the vigour of the new generation, the young bucks. The simplicities of the memorial and the nursery. What he did not see, or did not find to his taste, were all the awkward stages in between. The bending and twisting of trunks towards the light, the compromises wild trees have continuously to make between 'variety and grandeur', the diseases which may have effected the giant trees long before they were thrown down by epic storms, the fact that natural forests are not always awash, in all places and at all times, with thrusting young saplings – these are aspects of natural forests, too.

But Buxton got his way. Large numbers of pollards were cleared away, just as they were by the tidy-minded landscapers of the 18th century. Those that remained were never cut again, and began to shade out the regenerating saplings, as well as the ground-flora of the once much better-lit forest floor. It's for this he has incurred the wrath of modern conservationists, a pragmatic bunch for the most part, and more concerned about historic landscapes and biological diversity than airy-fairy concepts like 'naturalness'. They would have preferred the Forest to have been maintained in the state it was before the 1878 Act.

Edward North Buxton died in 1929, aged 83, but at the close of the Second World War his son was on the government committee charged with preparing a national conservation policy, and chaired by Julian Huxley and Arthur Tansley. The committee recommended that Epping Forest become a National Nature Reserve, and during the course of its work the younger Buxton confessed that his father had been wrong about the pollards: 'when pollarding was stopped the next 20 or 30 years produced close canopy conditions even more intense than in planted woods. It is probable that more species have been lost in the shrub and field layer to this natural reaction to artificial conditions than to the supposed inroads of an urban population.'

His contrition is understandable. Over the past one hundred years the regenerating trees have shaded out most of the Forest's undergrowth and ground plants. The plains and heaths, once kept open by the commoners' cattle, have been invaded by birch. It is, as a whole, a less scenically diverse place. But I can't share the pessimism of many ecologists – including Oliver Rackham – about its present state and likely future. It's still a wild and exciting wood, full of extraordinary trees. The pollards are smaller and less dramatic than those at Burnham Beeches, but have a distinctive Essex accent, sprouting perkily upwards rather than out because of the closeness of their neighbours. There are surviving giant coppice-stools, some on sites where they were noted in a survey of 1582. A few are those rare beings, pollarded coppice, where the regrowth has itself been lopped about 2 metres above the ground. The large areas of birch – a beautiful and fascinating tree often unfairly scorned because of its abundance – are being invaded by beech and hornbeam, and by oak where there's more light. In some areas the beech pollards are dying back prematurely (though this seems a very selective process, perhaps because of genetic susceptibilities to drought or fungal invasion), and as a result there is a huge variety of deadwood habitats for fungi and invertebrates.

On a simple biodiversity checklist, the species and variety of habitats at Epping have undoubtedly declined, as has its value as a museum of ancient land-use. But the Forest isn't an exquisitely planned landscape park. It's a living wood, which at some indeterminate time in the past was pickled by management, so that its existing species and structure were perpetuated on a roughly 15-year cycle. Now the processes of succession that were frozen when it was first extensively lopped – cycles taking many centuries to work themselves through – have been released again. Epping Forest is progressing naturally to beech-dominated high forest, in which the current big trees just happen to be unnatural pollards. This is a more monotonous habitat than the wood-pasture it once was, but with its own particular beauty, and the undeniable, vital fascination of naturalness – wildness – as a process.

What are the alternatives? None of them has any pretence at being economically based. It would be physically possible to clear most of the young birches, put in cattle to restore the plains, and resume pollarding the young trees – though this might not go down well with the people of Epping. At present what is happening is an inoffensive compromise, with patches of plain being kept open, and small-scale experimental pollarding resumed on younger trees. The Forest conservators have diplomatically restricted this to remote spots, and to the surrounds of car parks, where its justification on the grounds of safety are obvious. Such are the continuing squabbles about the identity of the 'natural' tree that a more widespread intervention would awaken the ghosts of the 18th and 19th centuries' hatred of 'mutilation'. And awaken, even amongst those well disposed to pollarding, an uneasy sense that this was not quite right, or real, on a par with 'distressing' furniture to give it an antique patina, or cosmetic surgery for the sake of prolonging interesting features into old age. Not bad aims in themselves, but not the trees' own ambitions.

IV

In France, the saving of an historic oak and beech wood-pasture came about from the intervention not of commoners or scientists, but local painters. The Fôret de Fontainebleau is a vast mixed forest (almost 140 square kilometres) south-east of Paris. Today it's chiefly an organised timber estate and one of the big leisure resorts of northern France. But in the 17th century it was both wilder and busier. Local peasants grazed cattle amongst ancient trees. The beech and oak were cut for firewood. Quarrymen and freelance stonecutters worked the forest's rubble of immense sandstone boulders, which were the source of the paving stones for Paris streets. There was even a small pumice-stone quarry.

In the 1660s there was an attempt to bring Fontainebleau, and all French woodland, into some kind of organised national system for the sake of the navy, but the scheme fizzled out. With almost 30 per cent of its land surface under wood, France scarcely needed to bring its insubordinate communes and haughty estates into line to find enough timber to fight a war. But as in Britain, the 18th and early 19th centuries brought greater strictures. Vast tracts of woodland were cleared for agriculture. During the turmoil of the Revolution there was the same looting of woods as in the English Civil War. By the mid-19th century there was still a good deal of woodland – maybe 20 per cent of the land – but rather less standing timber, and in 1824 the entire French forestry service was modernised. New techniques that had been developed in Germany, such as the use of conifers to replenish the soil, were imported – though without the Germans' mathematical rigour. The shelterwood system was introduced, and became the most commonly used forestry rotation in France. It's a gently manipulated version of the process by which natural woods – beechwoods especially – renew themselves. The trees in any stand are allowed to grow up until they begin producing seed. A proportion of these are harvested, creating gaps for regeneration and more light for the retained trees, which continue to produce seed. This process is repeated once or twice more until there is a final crop of the best timber trees, and several generations of youngsters growing up beneath them. The result, with its often dense mixtures of beeches of different age ranges, and broken here and there by tolerated groups of hornbeam and maple, looks authentically natural. But, because of a widespread belief among French foresters that the best age to harvest beeches is about 120 years, when they have made good timber but their seed-production is just beginning to decline, it looks oddly foreshortened too.

In Fontainebleau, the first conifers had been planted in 1786. In 1830, under the new regime, 2,430 extra hectares were put down to pine. But the Forest was already beginning to become the haunt of the Parisian artistic and middle classes, and a literate opposition to the 'industrialisation' of the forest began. A survey in 1852 revealed that some 70 per cent of the Forest was being used for 'industry', which meant any system of organised management. The eminent artist Théodore Rousseau, who had been painting the Forest since the 1830s, rallied his friends, and petitioned Napoleon III to protect the *forêt ancienne* from the debilitating cankers of commercial forestry and the tourist industry.

Rousseau was by then the undisputed doyen of the Barbizon School of painters, who had their focus in the village of Barbizon on the western edge of the Forest. They were a large and disparate group, from more orthodox landscapists such as Charles-François Daubigny, and Corot, through Gustave Courbet and Millet, whose uncompromising pictures of working peasants outraged the French bourgeoisie, to Monet, Seurat and Pissarro, whose work in the forest formed a bridge between the Barbizon School and the Impressionists. But they were united in an absolute commitment to nature, and to catching its energy in the open air, up close. They worked at the wood-face. Nicolae Grigorescu's portrait of *Andreescu à Barbizon* is an heroic emblem of what they stood for, a kind of credo. The artist is emerging from the forest like a hatching chrysalis. He has his umbrella and stick over his shoulder, but they have the look of the tools necessary for a long expedition, not the bric-à-brac of a day out sketching. He is looking directly out of the picture. But there's no third dimension to the painting, no distinction between foreground and background, just a shimmering plane of dappled green and yellow, a few blurred trunks. Andreescu is materialising from the greenwood. The artist is an outgrowth of nature, no more a passive witness than Corot's 'peasant'.

Rousseau's paintings of the forest are also outstanding for the way they eliminate any sense of a horizon, of a limit to the landscape. The forest is all-encompassing, the *fons et origo. The Forest of Fontainebleau, Morning* (1845–50) echoes Corot's picture in its great arch of trees, which frame a group of cattle in a swamp, and beyond them a pool of hazy dawn light, glowing at the heart of the picture. *Le Vieux dormoir du Bas-Bréau, forest of Fontainebleau* (1837–67) is a darker picture, the heart of an autumnal wood, in which the browsing cattle have the indistinct look of sandstone boulders. His enigmatic *Under the Beech Trees, Evening (The Parish Priest)* (1842–3) is the closest he comes to outright Impressionism. The priest is squatting with a book in his hands under a strange line of thin and twisted forest beeches. There's no sense of where he is. The trees rise from a vast and featurless plain of bracken-brown, their autumnal leaves picked out against a twilight blue sky. It's a portrait of a hermit in the wilderness.

The extended date-frames for these pictures are an indication of the unconventional way Rousseau worked. He developed a technique which wasn't the result of fleeting glimpses expressed in rapid brush-strokes, but of a sustained visual exploration, an intense gaze stretching over many years. His biographer Alfred Sesnier refers to what he calls Rousseau's 'deep knowedge', an intimacy with the landscape achieved by prolonged concentration. Instead of creating a landscape painting by the usual process, which involved preparatory sketches, tonal experiments, paper ideas, he did all this evolutionary work on the canvas itself. The paintings grew almost biologically – sometimes at much the same rate as their subjects, and with the inflexions and twists of fate of their pasts incorporated. They engrain the passage of time in the same way as the trees themselves.

At the same time as the Barbizon painters were in deep exploration of the meaning of the Forest, Paris was becoming dizzy with its scenic charms. Seduced partly by the rhapsodic canvases emerging from Barbizon, tourists flocked to the region, helped by new rail links. A 14th-century refuge for a real hermit was converted to a kiosk to sell cigarettes and beer. A new kind of guidebook appeared, suggesting paths that the curious visitor might follow amongst the great trees and fabulous rocks. The first was Etienne Janin's *Quatre Promenades dans la forêt* in 1837, followed by Claude-François Denecourt's *Guide de voyageur,* three years later. Denecourt was an obsessional hiker as well as a driven entrepreneur, and it was his determination that led to the current network of blue-arrowed pathways that criss-cross the Forest.

The local artists' view of the Forest hardly coincided with the manicured tourist honeypot they'd unwittingly encouraged it to become, and one writer, Emile Bernard, complained that it was acquiring the 'cultivated air of an English landscape garden'. But for the French government, the simultaneous rise of Fontainebleau as a commercial gold-mine and a centre of French artistic excellence couldn't have been handier, and Napoleon III readily agreed to Rousseau's 1851 petition. In an unprecedented move, he authorised the creation in 1853 of some 400 hectares of *Réserves artistiques,* groups of especially beguiling old trees and rocks, which were to remain uncut and unmanaged. It was the first nature reserve ever to be established for aesthetic reasons, and the first reserve of any kind in western Europe. (Czechslovakia had been creating forest reserves like Boubinsky Prales since the 1830s.) The reserved area was expanded to 1,000 hectares in 1861, and then into a full-scale *Série Artistique* covering one-tenth of the Forest (4,000 hectares) by 1904. But during the period of intense forestry modernisation in the 1960s, large numbers of the old trees were cleared away, victims of the now familiar forester's supersitition that young trees would never regenerate and grow unless humans removed their parents. *Dépérissement* – 'decline' – was looked on not as an entirely natural part of the woodland cycle but a kind of sickness or repressive force. In the more ecologically aware 1990s, the *Série Artistique* was formally abolished, and replaced with a series of *réserves biologiques* covering more than 4,000 hectares, of which nearly 2,000 were to remain entirely unmanaged.

But the bulk of the Forest is manipulated in some way. In one of the broad tracks there is a noticeboard, explaining that the prospect ahead was once the scene of one of the famous Barbizon School paintings: Jean-François Millet's *L'Allée aux Vaches*. A photograph of the painting is part of the display. The view hasn't changed much in a century and a half. The Forest authorities try to keep it as it is, but 'interventions are necessary from time to time for the sake of safety and the view'. Also for the sake of the next generation of trees, '*pour favoriser le développement des petits chênes*'. In a rerun of what happened at Epping, the old oaks are being sacrificed for the young.

V

One of the *réserves biologiques* is La Tillaie, an area of the Forest named after the lime trees that would have been dominant here in medieval times. I visited it one autumn, but couldn't 'see' it at first, expecting the old beeches to look as they do in Britain, squat and knobbly, icons of the Picturesque. But this was a zone that had not been cut for at least two centuries. There were no pollards. What swam into shape as I edged into the wood were the forms of authentic natural beeches, towering columns maybe 250 years old and up to 40 metres tall, as smooth and uncluttered as trees in mature plantations – except that here large numbers of them were in a state of serious *dépérissement*. Some had already lost most of their trunks from about 3 metres up, the result of invasion by – ironically for this painterly place – what is called artist's fungus, and by its relative *Fomes fomentarius*, which forms horny, ridged brackets like carthorse hooves, and usually brings on a terminal rot in mature trees. The brackets crash to the ground like meteors as the trees decay. On one tree, still producing leaves about 10 metres up, they rose up the trunk like stepping stones, weaving their way between black woodpecker nest-holes and insect borings. I've kept a photograph of this beech, soaring beyond the camera's reach, as an antidote to the pollard reflex, and a reminder that this is the way a natural ecosystem provides space for the next generation.

And the youngsters were behaving exactly as they must always have done in unmanaged forest. Around the dead trees there were new beeches of all ages. In one clearing they were growing up between the fallen trunks of four adults which had crashed down together. In another they were filling the space created by a fallen lime, one of the very few still growing here, and which was itself sprouting new poles from its prostrate trunk. One dead beech had fallen with extraordinary precision into the narrow fork of its neighbour, whose branches were already starting to grow in embracing curves around it. There was a beguiling entangledness about the place, the trees, young and old, growing through and around each other.

Later I discovered a wild medlar tree, boar rootlings, big globes of *Ramaria* fungus hanging high in the trees, like ochre coral. But no young oaks. There has been just one occasion in the past 150 years where the gaps were sufficiently big to start a new generation of oak. This is not their moment in the cycle of succession. But La Tillaie, with its straight uncut beeches and gaps full of young trees, may be the most authentically natural beechwood in northern Europe. Except for the fact that I was in it. As I walked out I saw a discreet notice announcing La Tillaie was a strict non-intervention zone, and the public were not allowed to enter. I had become an intervention myself, as I had, very much more aggressively, in Hardings Wood.

VI

Fontainebleau was at least saved as a wooded area. So were Frithsden, Epping, Burnham Beeches and the New Forest. Many beechwood commons, with lower public profiles and less spectacular trees, were not so lucky. Before the Metropolitan Commons Act many beech wood-pastures around big towns were cleared, for agriculture or housing development, especially in Surrey and Essex. Hainault Forest, next to Epping, was almost entirely destroyed. But in the Chilterns, the beechwoods as a whole had a remarkable upturn in their fortunes in the 19th century. Their long decline during the previous century, in the face of the increasing availability of cheap coal and the profits that could be made by growing wheat instead, went into abrupt reverse about 1790. No one is sure whether what happened next was led by consumers or growers, but in the decade that followed a huge demand erupted for Windsor chairs as a cheap and modish item of home furniture, especially in the mushrooming suburbs of the Midlands and the industrial North. The round legs, spindles and back supports were traditionally turned from beech (the seats were usually elm), and the wood merchants in the Chilterns realised they could be back in business. From Chesham in the north to Stoke Row in the south, chair-making became the fastest-growing occupation. In West Wycombe alone the number of turners and joiners rose from 18 in 1798, to 95 in 1841 and 431 in 1871. To meet their needs for slender, turnable poles, some new beech plantations were started – mostly on the sites of ancient woods that had been cleared for agriculture a hundred years before. But chiefly the surviving mixtures of coppice and high forest were put to work with a new kind of product in mind. The selection system – the simultaneous extraction of particular, chosen trees of all ages – moved up a gear. The bigger trees were harvested younger. The thinnings went to the chair-makers as well as for fuelwood. Some of the coppice was 'promoted' to high forest by cutting out all but the straightest leading stems.

I think this is what happened in one of my boyhood haunts, Hockeridge Wood, a mixed oak and beechwood on an acid ridge midway between Berkhamsted and Chesham. Browsing in the county record office, I once chanced on a few loose account sheets covering the years 1797–9. They recorded an immense sale of firewood faggots, a quantity which could only have been generated by coppicing the entire area of the wood in three years, as against the usual eight to twelve years. Many of the sales were to named individuals – Geary, Pudephatt, Redding – whose families still live in the area (a Mr Pudephatt taught us woodwork at school), and changes in the way woods were cut may have had an impact on domestic life as well as on local economics and ecology.

Most of the chair-makers were solitary home-workers, or were linked through an agent in loose co-operatives. Pubs were often the centres for groups of workers, as at the Bird in Hand in West Wycombe, and the Crooked Billet (a billet is a piece of cut firewood) in Stoke Row. But the most celebrated of the chair-workers were the bodgers, who worked inside the woods, cutting and turning the beech *in situ*. It's a word that's now acquired unfortunate connotations, but the original bodgers were immensely skilled workers, using a technology that was at least two thousand years old. The billets of wood, cut straight from a young beech trunk or coppice pole, or split from larger stems, were first roughly shaped with a draw-knife, then mounted in the twisted rope harness of a treadle-driven pole-lathe, spun at high speed and shaped with a hand-held chisel.

H. J. Massingham met one of the last working bodgers in the 1930s, and saw him as a romantic survivor of 'the changeless rhythms of country crafts'. 'Talking to that patient, soft-voiced, mild-spoken man', he wrote, 'and listening to his tale of resignation within the core of Nature's undying life, I could know the true, the timeless England at the moment of its last sigh. Yet the indestructible reality seemed to be in the tonal harmony between the skirr of the chisel and the murmuration of the wind in the high beechen tops rather than in the forces of the raging world where neither one nor the other is heard.'

Massingham had always suffered from a bad case of selective perception, and in reality bodging was nothing like as glamorous as he painted it. In the winter cutting season, the bodgers lived in 'hovels', built from branches and insulated with beech leaves and the shavings from their lathes. They were miserably paid, by the gross of legs, with the stretchers and spindles being 'included' in the fee. It was only the most fiercely independent who were able to survive when, inevitably, the shaping and turning became mechanised.

By the 1860s the trade to Australia and New Zealand was partly met by sending out flat-pack kits – the unexpected originals of the modern trade in do-it-yourself furniture. Soon, the seats of the chairs were also being manufactured from beech, instead of the traditional and more resilient elm, with embarrassing consequences for the more substantial purchasers. And with a ready supply of beech timber, the new factories began diversifying, and turning out floorboards, kitchen tables, paintbrush handles.

Mechanisation had an impact on the beechwoods, too. Trees were selectively thinned, so that the remainder would grow straighter and more uniformly. Increasingly the growing of the trees reflected the mass-production philosophy of the chair-making industry. By the 20th century, clear-felling, followed by replanting with nursery-grown stock (in the company of conifers, to 'pull the beeches up'), began to challenge the traditional Chiltern system of selection-felling and natural regeneration. Inexorably the woods began to move towards the monocultural, even-aged, deeply shaded places that are regarded as the archetype of beechwoods today.

But at least the beech presence survived, and with it large numbers of the ancient woods, even if they now grow in different forms. Even the new plantations make a contribution to the general woodiness of the region. And surrounded as they often are by so many semi-natural woods, they have already begun to be invaded by some of the Chilterns' ancient floral denizens. The planting of the timber beeches at Hardings Wood in about 1880 returned the wood to its old outlines, and sometime during the same decade a hundred years later, not long after we'd started work there, a sweet woodruff seed found its way across from the old part of the wood into the plantation, and sprang into flower.

Chapter Seven: The Beech Electric

At last the beech is fêted. Visionary artists and meticulous scientists prowl the English chalk-hills, looking for the source of the power of this odd-tree-out, this sensitive rebuff to the oak. Images and universal theories are developed, only to be turned upside down one night in the 1980s.

I

IT WAS, FOR most of my generation, the first forest of childhood. The dark shadow on the winter horizon, 'like a black reef in some still southern sea', the faces glimpsed in louring trunks. Then Mole stumbling over logs and root-holes, running in panic from the pattering behind him, a sound 'like sudden hail on the dry-leaf carpet spread around him'. The Wild Wood in *Wind in the Willows* – such a powerful presence in our cultural memory that it elided easily into 'wildwood' as the accepted term for the primeval forest – is, as spelt out by Kenneth Grahame, an archetypal beechwood. It could not, given the story that was to unfold in it, have been any other kind. Ernest Shepard's throat-catching illustration shows Mole as an insignificant bundle in the snow, gazing up at that unmistakable profile: the great buttressed roots, the smooth bole ever so slightly inclined towards him. Not threatening, but magisterially beyond him. It becomes his refuge, before he's rescued by Rat and taken to the comfort of Badger's catacomb of a den. The 'deep dark hollow of an old beech tree, which offered shelter, concealment – perhaps even safety, but who could tell?'.

The early years of the 20th century saw something of a reaction against the stuffiness of the Victorian era. It was hardly a confrontation, more a side-step, a polite no-thank-you to established religion, excessive masculinity, bourgeois conventions. It was echoed by changes in feeling about the natural world. There was a growing distaste for the 19th century's organised plantations. England's native woods came back into fashion, seen as deep shelter for old liberties as well as for curmudgeonly badgers. Amongst the *literati*, trees became one of the totems of this mood. But it was not the patrician, down-to-earth, practical oak that they focused on, but the sinuous, changeable, androgynous beech. In 1904 the back-to-the-land socialist Edward Carpenter had a vision of a tree as a conduit of energy:

> It was a beech, standing somewhat isolated, and still leafless in quite early spring. Suddenly I was aware of its skyward-reaching arms and upturned fingertips, as if some vivid life (or electricity) was streaming through them into the spaces of Heaven, and of its roots plunged in the earth and drawing the same energies from below. The day was quite still and there was no movement in the branches, but in that moment the tree was no longer a separate or separable organism, but a vast being ramifying far into space, sharing and uniting the life of earth and sky, and full of a most amazing activity.

This passage appeared in Carpenter's *Pagan and Christian Creeds,* a book which caught one of the intellectual nuances of the early 20th century, in which there was an explicit flirtation with paganism amongst English writers and artists. J. M. Barrie's *Peter Pan,* with its faery narrative and hinting title, appeared in 1904. Eleanor Farjeon dedicated a volume of poems to the Goatfoot. And in 1908 Kenneth Grahame published his immortal fable of four middle-class animals in search of the rural idyll, meeting Pan and fighting a class war in the Wild Wood *en route.*

As pagans go, Grahame was of the weekend variety. He preached a kind of gentle animism while holding down a job as Secretary to the Bank of England. The thought of orgies would have made him curl up, but in his apologetic, slightly precious way, he was a nostalgic ruralist, a romantic who hated the Machine Age and the tyrannies of 'Society' but craved the security of village order, a libertarian and hedonist who would never dream of Going Too Far. In his sketch 'The Rural Pan' (published in *Pagan Papers*) he envisages a reclusive, aesthetic nature god, a Scholar Gipsy who turns up in remote country pubs, telling stories in 'the musical Wessex or Mercian'. Even the Wild Wood itself was far from being some alien and primordial forest. In the story, Badger explains how the wood had 'planted itself and grown up' on the site of an old city. There are hints in the story that Badger's labyrinthine sett is part of the infrastructure, the cellars and old passageways of this lost metropolis. Arthur Rackham's illustration for this section of the book ('Crossing the hall, they passed down one of the principal tunnels') shows the sinews of beeches rooted in the top of a stone gateway, with the earnestly conversing Badger and Mole standing on a mosaic floor. 'It was all down, down, down gradually,' Badger is explaining, '– ruin and levelling and disappearance. Then it was all up, up, up, gradually, as seeds grew to saplings, and saplings to forest trees, and bramble and fern came creeping in to help.' As for its weasely inhabitants, he is magnanimous: 'It takes all sorts to make a world.' The wood itself was a neutral refuge, open to all comers. Grahame, like Mole and Rat, may not have especially liked wild woods, but they symbolised an alternative to the steeliness of the industrial world, something regenerative, indefinite, subtly beneficent. Badger's account of the origins of the Wild Wood was prescient. It prefigured, in its layperson's way, a developing theory of how woods began, and the way in which communities of trees succeeded each other, that was to engross ecologists for the rest of the century.

The inspiration for the Wild Wood was a real beechwood, Quarry Wood, which hangs on the steep slopes of the Chiltern scarp near Grahame's old home. He'd spent the first five years of his life amongst the dour granite landscapes around Loch Fyne in western Scotland. But on the death of his mother in 1864, he was sent to live with his maternal grandmother in the Berkshire village of Cookham Dean. The Mount was a one-time hunting lodge, full of twisting staircases and woody alcoves. The large garden had a ragged orchard and immense copper beeches, and blended at its rim with the wild woods of the southern Chilterns. Just a mile to the north was the River Thames. Grahame stayed there just two years before being moved on again. But the roominess, the hedonism of Cookham Dean, and the space it gave to his imagination, seemed to nurture him more than the austere and puritan habitat of his birthplace, and throughout his adult life he never strayed far from the warm and accommodating landscapes of the chalk country. Thirty years later, with a son much the same age as he had been himself in 1864, he moved back to the village. Young Alastair was taken on walks in the local woods and lanes, and doubtless introduced, through the lens of his father's imagination, to the mock Gothic horrors that chilled Mole's heart.

A hundred years on, the wood still has something of that

enchanted atmosphere. Walk up from the Thames and you clamber into a labyrinth of mossy banks and sunken wood-tracks and windthrown beeches. After the Great Storm, the chalk-pits left by wrenched-out roots were filled with thickets of deadly nightshade – an eerie echo of a pit that had obsessed and fascinated Grahame when he was a child. But he wasn't really a woodlander. He liked openness, lounging, the sun on his back. If he threw in his lot with anyone it was the 'river-bankers', and, to an extent, with Mole's relieved realisation, when he finally stepped out of the Wild Wood, that he was an animal of 'the frequented pasture, the lane of evening lingerings'. When Pan makes his hushed appearance in the chapter entitled 'The Piper at the Gates of Dawn' it's not in some torrid forest glade, ramping with goat's breath and snaky lianas, but on a tiny island in mid-river, draped with willow-herb and meadow-sweet. Pan himself is described, and pictured by Shepard, as a wry and kindly (albeit hairy) Henley oarsman, lounging on the bank and rippling his muscles.

In 1926, Sylvia Townsend Warner published her spirited debut novel *Lolly Willowes,* a story which makes beechwoods the setting for another story of escape. Lolly is a diffident woman in young middle age, who's endured twenty years of self-effacement as maiden aunt and house-keeper to her brother's insufferably stuffy family in London. Only in autumn does she have a respite, a momentary surge of inexplicable longing, in which she feels her mind 'groping after something that eluded her experience, a something that was shadowy and menacing, and yet in some way congenial'. In one such mood she buys herself a bunch of giant chrysanthemums. The shopkeeper wraps them in white paper, and adds a few sprays of beech leaves for decoration. Lolly is entranced. 'They smelt of woods, of dark rustling woods like the wood to whose edge she came so often in the country of her autumn imagination.' The sprays came from near Chenies, the shopkeeper tells her, in Buckinghamshire.

On her way home, 'holding the sprays of beech as though she were marching on Dunsinane', she calls in at a bookseller's and buys a guidebook and a map of the Chilterns. Within a few weeks she has deserted her family and moved into lodgings in the fictitious village of Great Mop. And wandering in the woods, she begins to find herself and lose her neediness. She throws away her map, goes night-walking, has a revelation about the nature of regeneration. She had thought of spring as a denial of winter, 'a green spear that thrust through a tyrant's rusty armour'. Now she saw it as something filial, 'gently unlacing the helm of the old warrior and comforting his rough cheek'. When her nephew Titus, a dilettante with artistic pretensions, pursues her to the village and smothers her with his unthinking appropriation of the whole landscape, *her* landscape, she has an epiphany in a field of weeds, screaming: 'No! You shan't get me. I won't go back.' Coming to the edge of the woods she hears the mutter of heavy foliage in reply: ' "No!" the wood seemed to say. "We will not let you go." ' In her despairing outburst she has, she soon discovers, entered into a pact with the Devil, a benign gentleman who, echoing Grahame's rural Pan, appears in the guise of a gardener. Lolly feels liberated, 'secure in his dangerous keeping', and the story ends with her planning to sleep out all night, 'wading through last year's leaves and this year's fern she would penetrate into a wood and burrow herself a bed'.

E. M. Forster's first short story, written in 1903, also features a fleeing from the straitjackets of convention courtesy of a woodland spirit. 'The Story of a Panic' describes the encounter between a party of middle-class English tourists and Pan in the chestnut woods of northern Italy. Eustace, an 'indescribably repellent' teenage boy in their care, is the most affected by the experience, and at the close of the story, now thoroughly feral, he leaps out of a window: 'he alighted in an olive tree, looking like a great white moth, and from the tree he slid on to the earth . . . and disappeared among the trees'. Forty years later Forster recalled how the story came to him while he was walking near Ravello: 'The story just rushed into my mind as if it had been waiting there.' 'The Story of a Panic' was the first of half a dozen stories written by Forster before the Great War, all of which feature allegories of liberation occurring in the woods. The next year, his inspiration was even more direct. He was walking in Greece, and 'found The Road from Colonus hung ready for me in a hollow tree' (a plane). In 'The Curate's Friend' a priest, unlucky in love, is given solace by a Faun, and, surrendering his faith during the encounter, gains communion with nature instead. 'That evening, for the first time, I heard the chalk downs singing to each other across the valleys, as they often do when the air is quiet and they have had a comfortable day. From my study window I could see the sunlit figure of the Faun, sitting before the beech copse as a man sits before his house.'

Forster was fond of chestnuts and planes, but beech, I suspect, was his familiar, and a beechwood is the far-from-neutral setting of 'Other Kingdom'. Harcourt Worters, a well-off businessman from the Chilterns, buys a wood – Other Kingdom Copse – as a gift for his fiancée Evelyn. She's overwhelmed – 'The joy of possession had turned her head' – and she attempts to 'dance' a beech. A high-spirited Irish girl, she gets it wrong at first, performing a birch, with her arms tight above her head. But then she 'twitched up her skirts so that for a moment they spread out in great horizontal layers, like the layers of a beech'. At first the gift of the wood strengthens the relationship between Harcourt and Evelyn. But then things start to go awry, along familiar Forsterian fault-lines. Harcourt wants to build an asphalt path from the house to the wood, put in a rustic bridge, fence the whole place off to deter the village boys from carving their messages on the trees. Evelyn is shocked. It's her wood now, she doesn't care for a bridge or path, doesn't want her seventy-eight carefully counted beech trees to be fenced in, and rather likes the tradition of inscribing romantic billets-doux on the barks. 'It's called the Fourth Time of Asking,' she protests. 'I don't want it to stop.' The touching history she then relates is so plausible, and so in tune with the traditions of beech-carving, that it's hard to believe it's purely a creation of Forster's. The local boys and girls 'cut their names and go away, and when the first child is born they come again and deepen the cuts. So for each child. That's how you know: the initials that go right through to the wood are the fathers and mothers of large families, and the scratches in the bark that soon close up are boys and girls who were never married at all.' That is the kind of relationship Evelyn wants, unfenced, written in wood not on a paper contract.

But she seems to relent. The wood is fenced, and 'tethered by a ribbon of asphalt to our front door'. But a freak storm brews up. The 'trees of Other Kingdom frothed like the sea', and a huge beech bough is blown clean up the path and onto the lawn. When the weather calms down, they go into the wood. Evelyn does her beech-dance, swaying about in a green skirt. 'She danced away from our society and our life, back, back through the centuries till houses and fences fell and the earth lay wild to the sun. Her garment was foliage upon her, the strength of her limbs as boughs.' Harcourt, 'entrammelled in love's ecstasy', chases her through the wood, until, suddenly and inexplicably, she disappears into the trees. Harcourt suspects she has eloped with his ward, her secret lover, and is outraged at this betrayal: 'I found her no better than a savage. I trained her, I educated her.' But he is mistaken. She has, in the traditional mythic way, simply metamorphosed into a beech tree.

Not long after he moved to Abinger Hammer, on the Surrey ridge of the North Downs, Forster bought a small local wood called Piney Copse. Most of its timber had been felled during the Great War. He didn't much like oaks, because of their over-patriotic associations, and he restocked it with other local natives – birch, beech, rowan, wild cherry – and some well-behaved incomers, too – sweet chestnut, a few conifers. According to his biographer, P. N. Furbank, Forster approved of the arrival of self-sown saplings amongst his plantings, and of trees which had grown 'only because it occurred to them to do so'. He was also gratified to have found an entrée into village life. He threw the wood open for an annual school picnic, and hung the trees with bags of sweets and bunches of bananas. In 1934 he devised and wrote 'The Abinger Pageant', which was staged in the open air with 'The Woodman' as narrator.

Eight years earlier he'd written a short essay entitled 'My Wood', in which he tries to convey what Piney Copse meant to him. It gives the impression of a man less inclined to sociability and woodland liberty than the writer of 'Other Kingdom' and 'The Abinger Pageant'. The style doesn't help. It's reminiscent of Kenneth Grahame at his most arch and affected, an attempt to cover up the embarrassment of feelings with a show of facetiousness. Much of the piece is taken up with misanthropic jokes about breaches in his boundaries. He's miffed that 'his' bird has deserted his wood and flown off into his neighbour's garden. He deplores the picking of 'his' blackberries by villagers, and dreams of a fence as secure as Harcourt's. But behind this mannered grumpiness I think he is trying to say something serious about the emotional impact of 'owning' a piece of self-willed England – a discomfiting confusion that I recognise from my own experience with Hardings Wood. Possessing Piney Copse, he says 'makes me feel heavy. Property does have this effect . . . [it also] makes its owner feel that he ought to do something to it. Yet he isn't sure what. Restlessness comes over him, a vague sense that he has a personality to express – the same sense, which without any vagueness, leads the artist to an act of creation.'

What made these writers' (and there were more, such as Edward Thomas) focus on the beech? It may have been partly a geographical accident. Beeches were the most distinctive tree in the areas where they lived, on the southern chalk-hills. It may have been an explicit rejection of the masculine, patriotic oak of the kind that Forster made. But the way they write about the tree, the vision of its quiet persistence, its upward-rising growth – Edward Carpenter's 'electricity' flowing through the 'skyward-reaching arms and upturned fingertips' – suggests they saw it as reflecting their own dreams. The whimsical and vicarious forays these writers made into paganism were more about the spirits rising in their own heads than in the woods beyond.

But for Paul Nash, the 20th century's most transcendental painter of trees, paganism meant rather more. In September 1911 he visited Wittenham Clumps, two small but conspicuous neolithic earthworks topped with beech groves near Wallingford on the Thames, and had a revelation. 'The country round about is marvellous,' he wrote to a friend, 'grey hollowed (*or* hallowed) hills crowned by old trees of Pan-ish places down by the river. Wonderful to think on. Full of strange enchantments. On every hand it seemed a beautiful legendary country haunted by old Gods long forgotten.' The Clumps were to figure in his paintings for the next forty years.

Nash had been painting trees since he enrolled in art classes at Chelsea Polytechnic in 1906, aged 17. He drew them in Hawk's Wood near Iver in Buckinghamshire, and later in Savernake Forest and the Forest of Dean. He was fascinated by the way natural forms and human artefacts developed when they were left to their own devices, and how one might change into the other. His feelings for the megalithic stone circles at Avebury reflected what Uvedale Price saw in his picturesque trees. The stones were 'In their wild state, so to speak . . . Their colouring and pattern, their patina of golden lichen, all enhanced their strange forms and mystical significance.' He first painted the Clumps – which had been planted with beech in the mid-18th century – in 1912, and commented later:

> I felt their importance long before I knew their history. They eclipsed the impression of all the early landscapes I knew. This, I am certain, was almost entirely due to their formal features rather than to any associative force . . . It was the look of them that told most, whether on sight or in memory. They were the Pyramids of my small world.

Nash's first portrait of the Clumps, *The Wood on the Hill* (1912), is a compelling image, and it's not hard to see how the mounds became his obsessional English equivalent of Cézanne's Mont Sainte-Victoire. As in all the pictures that followed, the hill, a fort, is seen from a distance. The corn that covers it has just been cut, and a pathway winds between the stooks up to the wood on the summit. The beech grove looks resplendent, a summer chaplet, and above it a soaring congregation of rooks seems to be drawing the whole landscape up into the clear air. The trees are leavenings, levitators, like Carpenter's electric beech, and the 'strange forms and mystical significance' of this Iron Age pyramid seem not so much enhanced as overwhelmed by their friskiness.

Nash's landscapes, and their constant fascination with the interplay between the natural and the artificial, always have this ambivalence for me, with nature sometimes seeming, in his word, to 'correct' the human-shaped and the inorganic, and sometimes seeming to be absorbed by them. After the First World War he visited the Chilterns often, and in 1929 painted his best-known picture, *Wood on the Downs*

– a study of the same beechwood at Ivinghoe where I'd had my own revelations of the interplay of tree and light in 1974. It's an indelible image, that quiff of dark foliage gelled into a curve that exactly echoes the shape of the downs behind. Yet the beeches are scarcely living trees at all. All details of leaf and intricate branching have been brushed out. Their trunks seem to have been dusted white by the patches of bare chalk in the foreground. It is as if they were outgrowths of chalk themselves. Beyond the purely formal rhythms of the composition, there seem to be deeper resonances. Has rock metamorphosed into beech – something which, in the literal world, it truly seems to do? Or are the trees already embryonic fossils, like the myriads of tiny sea-organisms whose skeletons formed the chalk 70 million years ago? History and destiny seem inextricable, as they are in the *The Wood on the Hill.*

The Chiltern landscape itself is full of such ironies and illusions. Between 1919 and 1921 Paul Nash often visited his brother John's cottage at Whiteleaf, a village on the scarp near Princes Risborough. In the beechwoods that rise above the village, there's a clearing which contains a huge white cross carved in the chalk. Its origins aren't known. H. J. Massingham, ever the romantic, believed that it was neolithic. The historian Arthur Mee, ever the patriot, thought that it had more likely been carved during the Restoration. Nash would have relished the uncertainty of its significance, but perhaps, even more, the fact that from its tip there is an unparalleled view of the Oxford plain, and on the far horizon, blurred by 20 miles of low-lying air, the twin mounds of Wittenham.

His last paintings of the Clumps were made from a marginally more intimate distance, the 12 miles that lay between them and Boars Hill, a village south of Oxford. In the early 1940s Nash stayed there at a friend's house, which also had a view of his totem trees. He was already ill with the lung problems that would kill him five years later, and used to content himself with looking at the beeches through binoculars. This remote, restricted view seemed to heighten their 'compelling magic' for him, and the series of voluptuous seasonal landscapes he painted between 1942 and '44 have the compact and foreshortened look of prospects squinted at through a lens. The beech clumps lour at the very centre of each painting, as perhaps they did at the focal point of his binoculars. Their complexion barely changes, and they are pictured as featureless brown tumps at both 'the Vernal Equinox' and 'the Summer Solstice'. They seem to be in a different visual and temporal space to the luxuriant vegetation in the foreground, completely absorbed by geological deep time. Bedrocks. They could be stone henges or rusting spacecraft from another era.

I wonder if Nash ever ventured *inside* the Clumps, or if he felt that kind of intimacy would undermine their distant magic. He might have found himself unexpectedly thrilled. On the eastern edge of one of the groves there is an ancient beech called the Poem Tree. In 1844, Joseph Tubb, from the nearby village of Warborough, followed the old graffitists' tradition and inscribed a poem around the tree, which was then about a hundred years old. It's a short (just 20 lines) but epic doggerel, and describes what had happened around the Clumps over the past two millennia: the bloody wars with the Danes, the Roman occupation, the collapse of the monasteries. Nash, a part of the scene as well as its portrayer, would have sympathised with its final lines on the impermanence of human ambition:

> Such is the course of time, the wreck which fate
> And awful doom award the earthly great.

The Poem Beech died between the wars, its death most likely hastened by the expanding perforations in its bark. The words are now illegible. But in 1965 the poem was traced by Dr Henry

Osmaston; and, laid out flat, with the letters stretched by a hundred years of bark growth, has the look of an archaic scroll, a piece of primitive calligraphy. Another human artifice brought down to size by the anarchic forces of nature.

Just before the Boars Hill series Nash had been working as an Air Ministry war artist, and seemed, in the fury of the Blitz, to find a more heroic role for nature. In the First World War, when he was first a front-line soldier fighting at Ypres, and later an official artist at Passchendaele, he found no hope or consolation whatever in its recuperative powers. *Vimy Ridge* (1917) and *The Menin Road* (1918) are pictures of utter desolation, landscapes of stripped and broken trees, lit up by searchlights and devoid of life, save for a few shadowy soldiers scurrying amongst the bomb craters, barely distinguishable from the sea of mud around them.

But in 1939, on the eve of war, he had a tree experience which seemed to reignite his youthful intimations of Pan. He'd been touring in the Malverns, and discovered two fallen elms in a field, one torn up by the roots, the other snapped off a few feet from the ground. They lunged out across the field, and Nash felt that, 'horizontally, they had assumed, or acquired, the personality of monsters'. He was spellbound. He sketched them, took photographs, wrote a short essay about the experience (called 'Monster Field', and published in a limited edition in his original handwritten text). These, he makes clear, were no ordinary dead trees: both were

> bleached to a ghostly pallor wherever the bark had broken and fallen away. At a distance, in sunlight, they looked, literally, dead white, but, at close range, their surfaces disclosed many inequalities of tone and subtle variety of ashen tints . . . Here and there the smooth bole, gouged by the inveterate beetle, let out a trickle of yellow dust which mingled with the red earth of the field.

They were transmogrified, existing – like the Wood on the Downs – in some middle ground between life and organic death. In their new life

> their roots and trunks formed throat and head, the uplifted arms had become great legs and hoofs outstretched in wild, mad career. What were they like? What did they most resemble in life? . . . We are not studying two fallen trees that look like animals, but two monster objects outside the plan of natural phenomena. What reference they have to life should not be considered in relation to their past – therein they are dead – they now excite our interest on another plane, they have 'passed on' as people say. These inanimate natural objects are alive in quite another world, but instead of being invisible like so many of that huge community . . . they are so much with us that I was able to photograph them in full sunlight.

He's suddenly reminded of Picasso's bulls, and begins to fantasise that his monsters might catch the foot-and-mouth disease that's smouldering in the district . . .

One year on, these wild visions of the transmutation of the inanimate take wing in his dramatic cameos of the war on the Home Front. Nature is absorbing the machine. A crashed bomber is sucked down into the corn. Fighters' vapour trails spiral like streams of mayflies against the sun. *Totes Meer* ('Dead Sea'), Nash's chilling portrayal of wrecked German aircraft sinking into the water, is far from convincingly dead. The grey hulks have the look of spawning fish, or molluscs emerging from a primordial sea. As for *Messerschmidt in Windsor Great Park* – an alien invader nose down in the ground, tailplane splayed in the air – it could be one of the Park's own goblin pollards.

II

At the same time that Nash was sitting on Whiteleaf Hill, gazing out at the distant smear of Wittenham Clumps, a Cambridge PhD student called Alexander Watt was peering at the ground in the woods close by, searching for beech saplings amongst the thorn scrub. There's no record of the two men ever meeting, but it would have made an intriguing seminar on the hill, fascinated as they both were with the rising-up of the beech. Nash was looking from a distance for signs of some unifying visual principle in the generation of life-forms from inert matter; Watt, microscopically examining the life-processes of the beech and how it became established. Both searching for patterns – 'Pattern and Process', as Watt titled one of his historic papers. He might well have titled his work on woodland (with a nod to his contemporary Rudyard Kipling's *Just So Stories*), 'How the Beechwoods Became', because it was concerned above all with the mechanisms of succession, how one type of wood naturally gave way to another, and how a species as vulnerable as the beech could ever get a root-hold in established woodland, and eventually become the dominant tree. To understand the mechanism of beech succession was one of the grails of early British ecologists, because they believed beechwood was the 'climax' vegetation in the south of the country, the stable, mature forest that, without human intervention, would eventually cover most of the land.

The idea that there might be such uniform and predictable processes in the development of woodland originated in America in the second half of the 19th century, when the philosophy of nature was heavily influenced by the Transcendentalists. They believed that the material world was the product of some ideal force. 'Nature is the incarnation of thought', wrote their guru, Ralph Waldo Emerson, 'The world is mind precipitated.' American natural scientists might have scoffed at the mysticism of this, but the sense of an irresistible upward development of living systems, the manifestation of some preordained pattern, permeates their thought at this time too. (Though in 1860, the most grounded of the Transcendentalists, Henry Thoreau, suggested a categorisation of woodland more concise and more rooted in the muddled realities of tree growth than was to appear for another century. 'Primitive Wood . . . Never been touched by civilised man' is at one end of his table, and 'Artificial woods . . . Those which have been set out or raised from seed, artificially', at the other.)

Succession was first recognised by H. C. Cowles, studying the vegetation of the dunes by Lake Michigan in 1899. There lime was the first tree coloniser, but was later displaced by maple and American beech. He recognised that climaxes were unstable, that they might never be reached, that changing conditions could destroy them. Frederic Clements, writing twenty years later, elaborated these ideas, but into a more rigid and complex structure. He had a semi-mystical belief in plant communities as a kind of super-organism, and thought that climax vegetation was the 'adult' community, 'the final form of a vegetation of a climatic region through a climatic period', the condition of greatest stability. His theory, like Cowles's, at least injected an element of dynamism into the static descriptions of plants and vegetation which had dominated ecology up till then. But it amounted to a belief in a kind of biological predestination, which seemed to pay no heed to the uniqueness in time and space of every community of trees, or to the intrinsic 'wild playfulness' of nature.

But it soon became the conventional wisdom in Britain, too. A. G. (later Sir Arthur) Tansley, Watt's supervisor at Cambridge, was the doyen of plant ecologists and populariser of the term 'ecosystem'. Tansley took up with Clements, and with his idea of an orderly succession of vegetation leading to a state of stability, the climax community:

> When a bare surface is presented to colonisation by plants these settle upon it and establish themselves in a more or less definite sequence . . . This *succession* of communities culminates in a *climax* community dominated usually by the largest and particularly by the tallest plants which can arrive on the area and can flourish under the particular conditions which it presents, thus bringing the vegetation into equilibrium with the whole of its environment.

His vision of Britain's mature prehistoric forest was of a barely broken mantle of trees, their mixture determined in any one place by soil and microclimate, the whole community slowly pulsing as ageing trees died and made space for their offspring, but fundamentally not changing its overall character. Beech would always be the main component of the climax forest where it was happy, because of its superior height and its ability to shade out all competitors. Succession was irresistible. It was the unconscious drive of nature at work.

If Clements wrote in the shadow of the Transcendentalists, Tansley was immersed in the ideas of Freud and the utopian socialists, which were buzzing amongst British intellectuals during the 1910s and '20s. There's no evidence that he was ever a member of the Communist Party, as were many of his Cambridge contemporaries. His social activities amongst field clubs and amateur naturalists suggest a rather liberal kind of radical – though his proposals for the introduction of a more dynamic ecology into university curricula were branded as 'Botanical Bolshevism' by the Regius Professor of Botany at Glasgow.

But in 1916, aged 45, a lecturer in Botany at Cambridge and Fellow of the Linnaean Society, Tansley had an intensely vivid dream that, he confessed later, changed his life. He was in the South African bush, faced by large numbers of Zulus with spears. 'I myself had a loaded rifle, but realised that I was quite unable to escape in face of the number of armed savages who blocked the way.' His wife appeared, dressed all in white, and drifted towards him 'unhindered by the savages'. Then he thinks he fired the rifle and woke up.

It was this dream that prompted his fascination with psychoanalysis. In 1922 he went into analysis with the master himself. Freud wrote to a colleague that 'I find a charming man in him, a nice kind of the English scientist. It might be a gain to win him over to our science at the loss of botany.' Freud did win, at least for a spell. In the spring of 1923, Tansley resigned his post at Cambridge, not returning to teaching until 1927, when he was elected Sherardian Professor of Botany at Oxford. In the four years in between he devoted himself to writing and lecturing on psychoanalysis.

He'd already made a major contribution to the popular literature. *The New Psychology and its Relation to Life* (1920) was a best-seller, going through eleven impressions in nine years, and eclipsing all his professional ecological writings at the time. It's an intriguing book, elegantly and clearly written, linking Freud's theories with the more pragmatic findings of social psychology in a way that made them accessible to lay readers. What stands out, nine decades later, is the prominence he gives to the idea of 'the herd' and 'herd instincts'. He means by this something uncontentious, simply all the constraints of human society upon the self. Yet there is an undertone of psychological determinism in the way he envisages the herd instinct as an irresistible unconscious drive operating through and modified by conscious cultural tuning. He has an old-fashioned belief in the perfectibility of Man, and the power of the psychologically mature mind to achieve this end: 'the ability of the mind to form ideals, is one of its most valuable powers. It may, indeed, be considered the highest faculty of the mind, without which human progress would be impossible.' In his conclusion there are echoes of the fears of savagery and disorder prefigured in his dream, and he writes that the loosening of the bonds of 'the old social order' by the Great War had 'set free many untamed instincts which have caused and are causing much damage and destruction . . . We must now look to a state [or did he mean State?] in which the individual must again be subordinated to the herd, to the national herd in the first place, but ultimately to the universal herd.' It sounds an authoritarian vision, yet he sometimes substituted the softer-edged 'brotherhood of man' for 'universal herd'.

In Tansley's great ecological study, *The British Islands and Their Vegetation*, published on the eve of war in 1939, this model of human society finds some echoes in his model of the succession of plant societies. The tree community is driven upwards by primitive forces, checked by constraints from its environment, but inevitably making stable progress towards a higher social order, a climax. Tansley admitted that major environmental disturbances, such as volcanoes, glaciations, human pressure, could interrupt or prevent succession to the climax. But these were exceptions, and he felt that the norm was sufficiently regular and orderly to be capable of expression in a schematic diagram of upward mobility: Pioneer Community to Pre-climax to Climatic Climax to, sometimes, Post-climax. Yet the Post-climax is the beginning not of collapse, but of some even higher state. Tansley barely refers to the degeneration of vegetation, or admits random disturbance into the picture.

The reality of succession is less orderly and predestined than this. Trees grow in communities and communicate chemically with each other, but they are not social organisms in the biological sense, not even the gregarious beech. Each individual and its family attempt to survive in the best way they can, not by being involved in some primitive group consciousness, or by bowing to environmental pressures, but by tacking, adapting, negotiating, competing. In the real world disturbance and compromise, not order, are the rule.

Tansley relied heavily on Alex Watt's findings in the woodland section of his seminal work. But he tended to subsume them in his grand theory, and missed one of their crucial messages: the disorderly, tortuous, intensely local, sometimes perverse character of succession. Watt, too, was fascinated by communities, the vegetational 'herds'. But his emphasis was the opposite of Tansley's. He saw the individual organism – and its unique responses to soil and weather – as the basis of the plant community, capable of giving development distinctive nuances and local accents. He was one of the first field-workers to follow the life-histories of individual trees.

Watt's decision to work on beechwoods for his doctorate was partly a result of the challenges this temperamental species poses to theories of succession, but also because the timber industry in the first decades of the 20th century was worried that beech was failing to regenerate as it once had. Between 1919 and 1925 he worked on a series of majestic papers which covered every stage of the beech's life, from seed to forest dominant. He was a lecturer in Forest Botany at the University of Aberdeen for the duration of his research, and began work in beech plantations close by. He put enclosures over sample plots of the local beechmast, some unprotected, others with different levels of protective wire-netting. In the unprotected plots every single nut lying on the surface was eaten by mice. Those lying under leaf litter did better, and those deliberately buried under the soil best of all. Worms helped, by burying the seed. So did a good mast-fall, which happened in the second year of the experiment.

In 1921 he began work in the natural beechwoods of southern England. On the South Downs in Sussex he found that access to *liquid* water, not just humidity, was essential for the beechnut's germination. This was more likely where the seeds were in moist leaf-litter, or had fallen into the tufted clumps of grass or cosseting plants. He found that all kinds of other factors influenced the beechling's fortunes: the compactness of the soil, which could vary from inch to inch; the frequency of frosts, which could not only kill the genetically more sensitive beechlings, but influence the migrations of mice from fields into the woods; even the activities of gamekeepers, whose indiscriminate slaughter of predators gave free rein to rodents.

The South Downs near Goodwood were especially suitable for studying the next stages of beech succession. The long agricultural depression, which had resumed after the Great War, meant that large areas of downland had been abandoned, and were reverting naturally to wood. Watt started what he called 'medical histories' for 200 natural, individual beechlings, from their establishment to their death. Snails and rabbits and a small leaf-cutting insect called *Typhlocyba douglasii* joined the army of remorseless predators. Whether a seedling survived was a lottery, but many plainly did, otherwise beechwoods would not exist.

Watt tracked their fortunes in the depths of the woods, and out on the downs. Every subtle difference in soil, wind exposure, sunlight, affected them. He measured the leaf-sizes of seedlings under different light intensities, and found that the young beeches were less tolerant of deep shade than was usually thought. He found that some individual seedlings sprouted laterally rather than upwards, seeking to reach beyond the shade of their parents.

Out on open grassland, beech had to wait its turn. The first stage in the succession to woodland was the establishment of scrub, clumps of hawthorn, juniper, dogwood bushes, themselves hugely varied, bent lopsided by persistent winds, browsed back by deer, opened up by rabbits. Then came the pioneer trees, ash, oak, birch, depending on conditions, and occasionally the odd beech. Sometimes a beech shot up first through a patch of thorn, and grew into a domed singleton. Sometimes, if grazing pressure was low, it colonised the grassland directly, without going through the scrub stage. Ash did better on the sheltered side of woods, oak where the soil was deeper.

Gradually beech seedlings established themselves near parent trees, though their advance was very slow, and Watt confessed he had no idea how they were able to travel the distances they obviously once had. But when the oak and ash were tall enough for the canopy to be more permeable to light, the young beeches began their ascent. At an age of about 100 years, they began to overtop all the other trees in the community, and eventually to shade them out. The first generation of beeches had broad crowns, growing without competition in the canopy. The next wave had necessarily narrower crowns, forming a second layer in the final structure. This was the climax, the fulfilled forest, and theoretically it should continue in this state indefinitely, renewing itself through 'gap-phase' regeneration. A single tree (or small cluster) dies back, or is windthrown. In this new patch of light, variations on the original succession are played out: through oak or birch where bramble colonises the space, through ash on chalky soils, with the beech seedlings clustering in narrow circles round their parent trees. Whichever route it took, beech would end up as the dominant tree in the canopy. The beech invasion seemed inexorable to Watt, its *modus operandi* reminding him, in an untypically memorable metaphor, 'of the Spanish Inquisition'.

But that was then. The southern beechwoods he studied had mostly been managed for centuries. They were only going through a period of colonisation of grassland because of the agricultural recession of the interwar years. The climate was in a quiet and unexceptional spell. Rabbits were at high population levels, and deer at low. Oak mildew had arrived, just, but not the nitrification of rain and ground-water.

What happened in those patches of the southern chalk could never be repeated in exactly the same way anywhere else or at any other time. Beech might always tend, briefly, to be the dominant tree in the south-east, but how it got there and how long it might be able to hang on were in the lap of the gods. The old riotous gods.

And that, in many ways, was Watt's point. His work is a celebration of oddity and divergence, a salute to particularity. He was no writer of elegant prose, but the intense detail of his writing has a kind of concrete poetry. His 'medical histories' are stories of real beeches. He doesn't talk about absolute rules, or exceptions to rules. The failures, the eccentric regenerations, the rogue saplings, the odd-beech-out – they are all part of the story. *Everything* that happens is significant. That is what nature is like.

At the dawn of the new millennium Franz Vera ushered a rather active bull into the comparatively settled china of succession theory. Going quite outside the field of the arguments of the previous century, he suggested that it was not some innate pattern of vegetational development which had driven the succession to woodland in Europe but grazing animals: wild oxen, boars, horses, elks. His thesis, put simply, was that former succession theories had written large herbivores out of the story. It was these which had created the patches of protective scrub through which tree seedlings could grow, and which played the crucial role in creating gaps for natural seedlings. When part of a wood died off (a big part, otherwise the theory wouldn't work) herbivores moved in and turned it into grassland. Their hooves helped disturb the ground and give seeds of woody plants a chance to become established. Patches of thorny scrub, which the animals didn't browse, would protect tree seedlings growing up inside it. The animals begat scrub which begat trees which begat the mature woodland. And this woodland was not the stable, closed forest

Tansley envisaged, but a kind of mobile savannah, kept open by ever hungry animals.

Many British ecologists, who have spent their lives watching animals eat woodland up, regard Vera as balefully as an evangelist on the doorstep. He's a driven man with an idea to sell. Although he had visited many sites, he wasn't a professional field-worker like Watt, and had worked up his thesis chiefly from a reading of a vast range of original sources – a selective reading, his critics say, of material that would support his case. And like all theorists, he has been influenced by the intellectual context in which he is working.

Vera had been an observer and then a supporter of the extraordinary developments at Oostvaardersplassen on the Dutch coast, where a large section of land claimed from the sea was abandoned in the 1970s. It turned rapidly into wild swampland, alive with birds, and the Dutch authorities decided to give it more nudges towards 'naturalness' by raising the level of the water-table, and introducing the kind of big herbivores that would have haunted such habitats in pre-agricultural times, with 'de-domesticated' cattle standing in for ox and auroch and Konik ponies for wild horses: 'analogues for the vanished wild beasts'. The experiment has been a huge success, with the animals creating a mosaic of vegetation – pools, marsh, grassland, reedbeds – amongst the burgeoning scrub. 'Now', Vera wrote, 'it appeared that in conditions analogous to untouched situations, nature developed as an analogy of those untouched situations.' For him this was not just an ecological revelation, but a political, even moral, parable, a story with which to challenge 'the prevailing conception of most nature conservationists in Europe that agriculture [meaning all systems of human management] is *essential* for maintaining nature'. Vera wasn't interested in a polite debate in academe. He was storming the barricades.

There's much circumstantial evidence to support his idea, both as a theory about the past, and a description of what might happen now. No ecologist would take Tansley's model of an unbroken shore-to-shore wildwood seriously any more. There was obviously, in prehistory, a multitude of areas which were open or had unstable vegetation: inland cliffs, flood-plains, swamps, deer grazing-lawns – the pollen records shows an abundance of non-woodland plants which must have had somewhere to live. But whether this meant that the whole of western Europe was an open savannah, the New Forest writ large, is another matter.

In modern Britain, too, it's common to see new trees shooting up through the rough scrub on any patch of rough or abandoned ground. But this tends to occur when grazing (or cutting) has been drastically reduced. Vera is only partially right about the protective qualities of thorny scrub. In winter, or at any time when grazing animals are hungry enough, they browse hawthorn, gorse, even holly, down to the ground. Except perhaps for one or two sites in the New Forest, there is nowhere in Britain where one could witness the 'degeneration phase' of his model – a large area of dead trees in a wood turned to grassland by grazing animals and only then able to evolve into high forest. Even after the storm in 1987, the considerable gaps opened up in native woods were rapidly colonised by natural tree seedlings without the help of so much as a single auroch.

There are too many exceptions to Vera's thesis to make it acceptable as a universal model of woodland succession. The shame is that he didn't need to give it such a grand and single-minded purpose. Herbivores were obviously a key component in shaping habitats in prehistory, and have shown that they could be again. They produce wonderful diversity in plant communities, which are both picturesque in the precise historical sense, and ecologically rich. They are an essential part of any truly natural ecosystem (though so are the vanished big predators which ate them, on which inconvenience Vera does not dwell). The developments at Oostvaardersplassen, which Vera helped to foster, have become an inspiration for conservationists across Europe, and are driving the big wetland restoration projects that are springing up throughout eastern England, and large-scale woodland restoration too. Herbivores are one of the factors which influence the dynamics of vegetation, but maybe aren't regenerating woodland's magic wand.

III

Those who are involved in woods all have their different hopes for succession: that it will prove theories, create wealth, deliver 'true' or 'natural' woodland. For John Workman, distinguished forester and one-time adviser to the National Trust on woodland policy, it was the hope that in his own Cotswold woods it would generate a continuous supply of high-quality timber, and a beautiful and ecologically rich landscape at the same time. And so it had been up until the late 1970s, when things started to go wrong.

I first got to know John in the 1980s, when I was serving a spell on the Nature Conservancy Council. He'd been generous enough to visit Hardings Wood from time to time, pass on advice and kindly comments on what we were doing, even lend a hand with stacking wood. We had a running joke about the effectiveness of grey squirrels as pollarders. His family had been in the wood business for generations, first as wood-mill owners, then as tree growers. They bought what would come to be called Workman's Wood, a large semi-natural beechwood near Stroud, at the beginning of the 20th century. This is John's legacy, and he has a family pride in it as well as a forester's.

I went down to see Workman's Wood one mid-November day, with the leaves still on the trees. John is bothered by arthritis, so we go for a tour in his Land Rover, and once we're on tracks that he's followed for the past fifty years, he becomes as nimble as a rally driver. He tacks down steep banks, knows all the corner-lines, and keeps up a running commentary on the beech movie that's unrolling past us. We rattle through the remains of a deer park, past old field systems edged with the ruins of drystone walls, both full of invading ash and beech trees. Then a one-time coppice, where the stag's-horn regrowth is snaking up to the canopy. I'm amazed at its wild shapes, but John points out the old paradox: the trees only have this unkempt profile because they were once cut. His favourites have always been the big, clear-stemmed timber beeches. There's a straggle of them on the slope to the right of us, one side of each trunk gilded by the setting sun and striped with branch shadows. They're delectable trees, columns of shot satin. But many have gone, 50 per cent, John reckons, killed during the baking summers of 1975 and '76. This wasn't just a consequence of the drought; it was the heat, too. Transpiration was so rapid in the high temperatures that the trees became seriously dehydrated in a matter of days. Some of those that survived never recovered the vigour a forester expects. John points out what he calls their 'sear' twigs, a sharp, foreshortened fretwork against the sky, not flexible and sweeping as they should be. He thinks their root-systems may have been permanently damaged in the fast-draining, rocky soil. He knows the moods and idiosyncrasies of every bit of this wood.

We can see what the beeches are growing in as we go round a spiralling series of bends, cut into the side of the hill so that the underlying rock strata are visible. There's only an inch or so of anything recognisable as soil. Below that are just layers of limestone clitter. John says they have to use pickaxes when they're planting trees. The beeches are regenerating, some of them under cover of ash, but he's not impressed by their growth, and doubts that they would have survived at all if he hadn't cut out nearby trees, to give the seedlings more light and root-space. And most of the current regrowth is not getting beyond a foot or so because of the pressure of deer.

John's cottage is in darkness when we get back, which doesn't help his sombre mood. The power lines go down regularly in this remote district. A few weeks before it had been a hapless barn owl, found frizzled under the wires. He never knows at the time, he complains, whether it's a terrorist bomb or one of his own trees, and I'm not at all sure which he would regard as worst. We sit in the darkness, talking about forest futures. He has a tinge of sadness about the fate of his woods, even though they are now a National Nature Reserve. He is a forester to his soul, and laments what he sees to be the future of woods as playgrounds, not the skill-based structures he has spent his life with. And he worries about the future of the native beech tree itself, racked by storm and drought, munched by forest animals, undercut in the market-place by cheaper foreign timber, and, he feels, threatened by the current laissez-faire, do-little policy in woodland conservation. Do *something*, he argues back.

George Peterken, one-time chief woodland scientist with English Nature, was led by the hope that he might witness the processes of succession in a way quite unhindered by human interference. Since the 1970s he's been observing the evolution of Lady Park Wood in the Wye Valley. The wood is very diverse with a good deal of beech, but also sessile oak, wych elm, ash and our two native lime species. In 1944, it was designated as a reserve to study the behaviour of natural woodland, and no substantial management has been carried out there for more than half a century. It was of course, far from virgin woodland at the time. It had been a coppice during the 17th century, a putative beech high forest in the early years of the 20th century, and had lost two-thirds of its standing timber during the Second World War. What the non-intervention designation meant was that it would be 'natural' *from then on.* George, who has a very lateral mind, has invented an original way of cutting through the vagueness and subjectivity of concepts of 'the natural', and their perennial linking with notions of some return to an earlier, ideal state. He divides naturalness into five types, including Original-naturalness, the un-recreatable state which existed before people became a significant ecological influence; and Future-naturalness, the state which would develop if human influences were completely removed, and which accepts existing compromises such as climate change and colonisation by introduced species as part of the baggage of history. Lady Park was intended to become a traveller on the road to some future naturalness, and the general assumption was that it would end up, by the iron rules of succession, dominated by beech.

What actually happened could scarcely have been predicted even by Watt at his most open-minded. Over two decades, the wood went through a series of exceptional and unpredictable disturbances. In 1975 and '76 two summers of extreme drought hit the beeches badly. Many of the larger trees were killed outright, and even the undamaged ones virtually stopped growing until the mid-1980s. In 1983 there was a furious season of beech-debarking by grey squirrels. George's colleague Eddy Mountford has measured this meticulously. By 1993 the squirrels had damaged 50 per cent of all the beeches, and in the vast majority of cases – more than 80 per cent – had chosen stems between 10 and 25 centimetres in diameter, reducing potential forest trees to mangled shrubs. In 1984, a late spring snowfall bent many of the thinner ashes and limes to the ground. The ash died, but the prostrate limes took root, and began sending up new stems. In 1985, a plague of bank voles killed almost all the beech saplings that were scattered about in the birchy areas. George had watched them in their feeding frenzy, and said that they were in such huge numbers that the ground appeared to be trembling. By the 1990s, after yet more winds, frosts and infestations, any notion that Lady Park Wood would make a steady and comfortable progress towards beech dominance was looking absurd. The wood had turned into a mosaic of all kinds of tree communities: ash and lime stands, drifts of birch, some beech and oak high forest, and more open areas with aspen clones and overgrown hazel coppice.

This was much how the wood looked when George and I walked round it in the winter of 2006. The beeches killed by the drought in 1976 had almost entirely rotted away, though there were still standing snags from later years. There was no visible regeneration at all, chiefly because of deer – except in one small area. There is a small cavern in the wood, and the Health and Safety Executive had insisted that it be fenced off. Inside this small enclosure – a deterrent to deer as well as humans – there is prolific beech regeneration, which poses an exquisite philosophical conundrum. Is it 'natural' to artificially exclude, for reasons which have nothing to do with conservation, a herbivore whose population levels are unnaturally high in order to facilitate natural regeneration? George believes that even his clear categorisation of naturalness is now thoroughly compromised by a proliferation of wild cards. Where on his naturalness scale would one place, for example, a deliberately planted native beech and a self-sown alien sycamore? Or the reintroduction of a locally extinct native like the black poplar, or the natural drift of Scots pine (native in England eight thousand years ago) out of its plantations? Is it origin or process that determines their status? Like John Workman, he thinks that in much of the landscape we should 'do something' to sustain and rebuild a recognisable, accessible, usable woodland estate. But not everywhere. The unmanaged reserve is still a place of fascination and wildness, even if its purity is a figment in the modern world. He has discovered a deliciously pertinent comment by Freud – who seems to buzz like a wasp around the certainties of ecological science: 'The creation of the mental domain of phantasy', Freud writes in *Introductory Lectures on Psychoanalysis,*

> has a complete counterpart in the establishment of 'reservations' and 'nature-parks' in places where the inroads of agriculture, traffic or industry threaten to change the original face of the earth into something unrecognisable. The 'reservation' is to maintain the old condition of things which has been regretfully sacrificed to necessity everywhere else . . . The mental realm of phantasy is also such a reservation from the encroaches of [reality].

As for me – well, I'm not sure what my expectations were of the patch of Hardings Wood where we'd felled the beeches in the early 1990s. In the sizeable clearing we'd made, I rather hoped that something would grow back, but preferably *not* beeches. I felt there were enough of those in the plantation already. Things did grow back, very rapidly, and I was heartened by the simple mass of green seedlings. So were our volunteer helpers, many of whom were still wedded to the belief that trees would not grow unless planted by humans.

But twelve years on, inspired by Watt's example, I wanted to take a more meticulous look, a tree-by-tree survey rather than a scenic overview. It was late April, and there were buzzards circling above the wood as I tramped up the track we'd made. They were nest-building in one of the plantation beeches, back in the parish after an absence of maybe two hundred years. Their return had nothing whatever to do with our tinkerings in the wood, but it was something I could never have dreamed of when I bought Hardings, and I felt exalted and ridiculously proud, as if they were a benediction on our labours.

The clearing, surrounded by quite large beech and a few ash, was obviously thick with new trees, despite the fact that grey squirrels and deer were energetic feeders in the wood. It was an elongated space, more oval than circular, and gently sloping from west to east. I paced out a rectangle of 2,500 square metres, and then measured every individual inside it. There were 139 of 14 species, almost half of which (65) were ash. The biggest ashes averaged 25 cm in girth, and were already 3 to 4 metres tall. Of a similar size were 4 cherries, 3 goat willows and 1 field maple. A shrubby layer between 0.5 and 1 metre tall included 27 hawthorns, 11 hornbeam seedlings, 5 elders, 4 spindles, 3 blackthorns and 1 each of holly, rowan and silver birch. There were also 11 hazels, belying the notion that nut-scrumping by squirrels prevented their regenerating from seed. When I drew a rough plan of the regeneration in my notebook, the first stage in succession, it did not look at all Wattian. Instead of a circular pattern, with the highest density of species in the centre, it was a pyramid, with the larger ashes at the higher westerly apex, and the vast bulk of the shrubby species crowded down near the lower baseline – which was also closest to the biggest variety of seed trees. In this throng were just 3 beech saplings about a metre tall. None of the felled beech trees, which had been cut at ground level, had sprouted again. I was vaguely disappointed about the beech, though this was the result I'd dreamed of all those years ago. This was natural regeneration proceeding in its own time and in its own way, despite having been initiated by an unnatural event – the close-shaving of a fifth-hectare of trees. More and more I'm inclined to view 'naturalness' not as a state, a place in freeze-frame, but as a process, a behavioural language, a movie. Naturalness is whatever occurs *between* human interventions.

Workman's, Lady Park and Hardings are three beechwoods showing different and often unexpected patterns of regeneration and succession. But none of them had been through a large-scale, sudden natural intervention, of the kind that over a long period of time is a regular experience for woods. Great storms are one such event, but the most bizarre succession to beech in Britain happened as a consequence of a landslide.

The Axmouth-Lyme Regis Undercliff, on the south coast of Dorset, barely existed before the 19th century. It was an area of rough cliff-top grassland and a little arable, which had experienced a few minor landslips. Then, at the very end of 1839, an enormous slice of inland cliff simply collapsed. A huge chalk floe slid away towards the sea, leaving a chasm into which the next section of the cliff collapsed. On top of it was a sizeable wheatfield, already carrying its spring shoots. It fell more or less intact, the right way up, and on 25 August the following year it was ceremonially reaped. The whole event became a cause of wonder and foreboding in the district. Ten thousand spectators gathered for the harvest, with the reapers led by young women, wearing brooches in the form of silver sickles. It was a celebration of a world-turned-upside-down, a wake for the passing-over of the domestic into the wild. In William Dawson Turner's water-colour the wheatfield stands in the centre of a vast amphitheatre of jagged chalk crags. There are people walking and picnicking, and on top of the cliffs a small group hoisting a Union Jack. The south of England still heroic in defeat.

For the next few years the Undercliff became a serious tourist attraction. Local farmers charged visitors sixpence a time to walk the paths. Francis Palgrave, the famous anthologist, and Tennyson (on the edge again) walked it in August 1867. Then came yet more landslides, slumps of clay, falls of great limestone boulders. Scrub began to invade the chaotic mass of rock and water, followed by the first pioneer trees, ash, maple and hazel, wreathed by lianas of old-man's-beard and ivy. But the land above the cliff carried mature beech hedges, and these also tumbled down. Many, like the wheatfield, must have come down the right way up and taken root where they landed.

In the heart of the Undercliff there is a group of enormous beeches which have the look of 200- or 300-year-old pollards. But they may be 'natural' pollards, broken in their fall and sprouting from their splintered trunks. Their seedlings have already started to colonise the surrounding vegetation, revelling in a climate of southern warmth and Atlantic moisture. The Undercliff beeches' future looks bright. They may not become the dominant tree in such a shape-shifting habitat, but they will take their place in the woodland community.

In 150 years the Undercliff has become one of the wildest woods in England, and probably the only one to which the coastal rescue services are regularly called out to search for lost walkers. John Fowles, who lived in Lyme, used the Undercliff as a key setting in his novel *The French Lieutenant's Woman.* It's portrayed as an amoral, turbulent backcloth against which the ill-fated Victorian lovers play out their own futures – though with its beeches rather more prominent than they could possibly have been in the 1860s. 'It looks', Fowles wrote elsewhere, 'almost as the world might have been if man had not evolved, so pure, so unspoilt, so untouched it is scarcely credible, so unaccustomed that on occasions its solitudes may feel faintly eery.'

IV

And then it happened. The Event of 16 October 1987. Compared to this great storm, which threw down 15 million existing trees and started the upsurge of a hundred times that number, the collapse of the Undercliff was a mere local ripple. Right from the beginning it was clear that it was going to be, above all else, a cataclysm of trees. I'd been woken about 3 a.m. by a surfeit of sensory oddities – screaming wind, unseasonal warmth, a premonition of dawn in the quality of the light. Even through closed curtains, the night outside looked *green*. Outside, the air was a whirligig of blown leaves, a stream of living ticker-tape. Every quarter of the horizon was lit up by flashes – not lightning, as I thought at the time, but power-lines short-circuiting. It was still blowing hard at 8 a.m., but the only casualty in the garden was one of the lilacs my parents had planted in the 1930s. The news told a different story, relaying a chaotic jumble of rumours and reports: crushed cars, massive electricity failures, the tumbling, they said, of most of the trees in south-east England.

My wood! What had happened to it? I sped up, but with mixed feelings, not sure what I was most fearful of, finding it flattened, or entirely untouched by the wild energies of the storm. The wind was still gusting at about 50 mph when I arrived. The blizzard of ash leaves was like waterweed in a stream. But almost all Hardings' trees had survived. One hornbeam had had its root-plate half-wrenched out of the ground. A few thin birches had been thrown into some thorn scrub. The brambles, strewn with shredded foliage, were standing on end, as if they'd been electrocuted. But not a single beech had lost so much as a branch. Capriciousness was to prove one of the features of this storm, which had picked out individual trees, houses, whole woods seemingly at random. And its centre had passed a little to the east of the Chilterns, which were to get their comeuppance in the next great gale in January 1990.

A picture of what had happened in 1987 began to take shape during the morning. What had struck southern England wasn't strictly a hurricane, but an extreme storm which had rampaged across western Europe with a violence not seen for three hundred years. The remains of Florida's Hurricane Floyd had been hurtling cold air across the Atlantic towards a high-pressure zone settled over Europe. The low-pressure jet-stream sucked warmer air towards itself as it tracked north-east, and behind it an eddying snake of turbulent atmosphere rushed between the Bay of Biscay and the North Sea. The temperature rocketed – more than 10°C in a few minutes. By 3 a.m. the average wind-speed had reached 70 mph, and wasn't to drop significantly till 6 a.m. Unbeknownst to Britain the continent had already been badly hit. In Portugal the area round the River Lima suffered its worst flood in a hundred years. In western France, a 25-mile band stretching across Brittany and part of Normandy had been virtually wiped clean of timber trees – 7 million cubic metres, twice the loss in England.

It was lucky that the storm hit England in the middle of the night. The ferocity of the wind – which was sucking matter into its path as well as simply blowing it flat – would have caused carnage if more humans had been about. Despite the widespread failure of power-lines and telephone cables, stories began to trickle through. There had been fatalities, mainly people crushed in their cars by falling trees. The caravan settlement at Canvey Island, one of the victims of the storm-surge flood in 1953, had been wrecked again. In Brighton, the streets were full of splintered glass as the gale flounced through, sucking out shop windows and bringing down many of the city's lovingly protected elms – then whipping one of the Pavilion's stone finials from its plinth and dropping it through the ceiling. In its last hours, the storm seemed to take a fancy to churches. Stained-glass windows were blown out. All that was left of Cransford Baptist Chapel in Suffolk was a single wall, and rows of pews standing under the sky. No fewer than three trees uprooted in Thelnetham Churchyard plunged through the church's roof. In churchyards all along the storm's track, pale human bones poked through the root wreckage to remind people, if they needed it, of the continuous existence of mortality in daily life. And down in Sevenoaks, Kent, where seven trees had been planted in 1902 to continue an oak presence that had given the town its name in the Middle Ages, just one was left upright.

Despite the loss of life, and material damage which was eventually reckoned to have cost £2 billion, the 1987 hurricane was perceived and remembered above all else as a catastrophe of trees. The immense cull lasted just a few hours, but it took with it all kinds of complacency about the immemorial virtues of the classic English landscape. And striking where and when it did, the storm was inevitably a cataclysm of beeches. In the great arc of chalk country that swings across south-east England – the Hampshire Hangers, the South Downs, the landscape parks of Sussex, the Kentish Weald, the Chilterns – beech is one of the defining trees. And at that moment it was, as it has always been, a tree of catastrophe. Heavy-leafed, shallow-rooted, it stood in soil saturated by months of heavy rain and, in the kind of military metaphor that became much favoured during the crisis, went down like hapless infantry before a cavalry charge.

Two days after the storm, I drove down from the Chilterns to the Kent/Sussex borderlands that were the epicentre of the damage. I'd heard an up-welling of emotions about trees in the previously indifferent English that surprised me, but also an anger about the damage that I didn't really understand. Someone had gone on television and said that the landscapes of southern England had been ruined for ever. He seemed very certain about what gave identity to a landscape, and to have no trust whatever in nature's powers of recovery, and I wasn't sure I believed him. I needed to see for myself.

The weather was kind, an interlude of gentle milky sunshine amidst the days of torrential rain that followed the storm. As I travelled south, it was hard to see much damage beyond the occasional toppled hedge-tree, and some crew-cut woods. The real maelstrom began imperceptibly, the way snow cover does as you rise up a hill. A few limbs in the road at first, then sudden tongues of flattened beeches at the edges of woods. I tacked through Windsor Forest, the nearest tract of really old trees. The timber trees in the plantations had been snapped and scattered, but the oak pollards in the Great Park had almost entirely escaped. Contrary to conventional forestry wisdom, great age and internal rottenness don't make trees vulnerable, but actively protect them. Most of the Windsor veterans were squat and hollow, with small branch areas and low centres of gravity, which made them very stable in high winds.

I dogged the track of the storm, seeing extraordinary sights: beeches with their tops twisted off, as if they had been hit by miniature tornadoes; beeches blown clean out of the ground, with their vast root-holes already full of water; whole plantations of conifers neatly snapped about 1.5 metres above the ground. Further south, what was most striking was the apparent randomness of the damage. I saw trees on the windward side of the road barely touched, and whole rows on the leeward, just a few metres away, completely flattened. I'd pass through a mile or so of country where the leaves weren't even scorched by the wind, then hundreds of yards of intense devastation. Some woods had their perimeter trees intact, but great gaps gouged out of their centres. Trees on the edges of woods have more extensive root-systems, growing out into the fields, and this makes them harder to topple. In general it looked as if the very old and the fairly young were least affected, with plantation trees and large-crowned, full-sailed 100-year-olds the typical victims, if there was such a thing in this chaotic cull.

When I reached north-east Sussex, there were places where every other tree seemed to be down. I edged the car along by-roads carpeted with a compost of frayed leaves, dodging lianas of fallen power-lines and telephone cables that were often only inches from the car window. Already the lanes were lined with tilting boles and sawn trunks, and beyond the thickets was the constant whine of chainsaws, as marooned householders tried to cut their way back into the outside world. The whiteness of the broken trunks and dislocated branches was astounding. But it's the smell that has stayed with me, a cocktail of warm sap from the split wood and the tonic, almost vermouth-like aroma of myriads of crushed beech leaves. I have to confess that, entirely selfishly, I found the whole experience intoxicating. The whole landscape seemed sprung, as full of latent renaissance as Nash's Second World War paintings.

But it was not pleasant for most people, with trees becoming agents of destruction as well as victims. One crushed a woman to death in her bed in Chatham, the only indoor fatality from falling trees. The sculptor Anthony Gormley's studio and the artworks it contained were ruined, and he responded in a way that was not quite in kilter with the general opinion. 'I regard the whole thing being in a sense nature pruning the works of man,' he told a *Times* journalist. 'There are times when I feel it was strangely appropriate . . . One whole aspect of my work has been to reposition man within a kind of elemental context.'

Out in the countryside, the elements too seemed to be suggesting a different kind of aesthetic. The elegant landscape parks of the Garden of England were comprehensively rearranged. Capability Brown's constructions at Sheffield Park and Petworth in Sussex were turned into free-form jungles. The grounds of Scotney Castle, developed into a pleasure ground of gentle Picturesqueness by generations of the Hussey family, became a wilderness in just three hours. The grove of beeches on top of the neolithic earthwork at Chanctonbury Ring, planted in 1760 in a similar manner to Wittenham Clumps, lost three-quarters of its trees. Some of the natural beechwoods, perhaps more securely rooted, escaped. Epping Forest lost no more than a few hundred trees. Selborne Hanger, on a downward, north-east-facing slope, was scarcely touched, though the more exposed slope of Noar Hill, just a mile to the south, lost about half its beeches. When I went down to Selborne, I found villagers talking with thrilled awe about the fifty beeches that had blocked a lane below another hanging wood. They'd had to be shifted by dynamite, which blew the road surface to pieces too. There was a strange mood in the air across the storm-struck zone, the sadness of loss entwined with the exhilaration of new opportunities. At Kew Gardens, where a Tree of Heaven had mischievously crashed onto the roof of William IV's Temple, they were talking of holding a memorial service 'for the Fallen'. In central London it was the planes that were being grieved over, and on their parchment-like leaves you could taste the salt that had been blown in from the Channel. A bitter aftertaste. In the week after the storm I watched adults wandering through the parks and squares in slow motion, lost in a state of incomprehension at what had happened to their home landscapes. And I saw their children having the time of their lives crawling about these vast natural climbing frames.

The seismic changes that the storm was to bring about in our cultural attitudes towards trees were also beginning to rumble. George Hill, in his reportage account of the storm, *Hurricane Force*, argues that it 'enforced a sharp, and one hopes lasting, change in our perception of the fact that trees in the landscape do not just happen to be there. They need planning, enouragement, investment . . . .' – a curious conclusion given how much damage had been done to those trees that *had* been planned and planted, and how many that 'just happened to be there' had got off scot-free. The real changes in perception were more complex than that, involving every emotion from anger to exultancy.

Bitterness came first. The Tree Council issued a press statement, an extraordinary solecism which seemed to place the republic of trees entirely inside the kingdom of man: 'Trees are at great danger from nature', it announced – as if the storm had been an actively malign force, beyond any normality – and declared that unless 'funds are made available and positive encouragement given to owners to restore these woods . . . they will revert to scrub and never recover'. Shades of the paranoia in the Chilterns in the 1970s. The National Trust held an art exhibition at Petworth, and prefaced their catalogue with an apocalyptic message: 'The Great Storm desecrated the past and betrayed the future.' Distinguished artists, including Derek Hill, Adrian Berg and David Gentleman, painted scenes of devastation from Scotney Castle to Cliveden, but as is so often the case in the art of the natural, could not always mask their fascination with the enormity and visual exuberance of this huge act of transformation.

But the storm had happened, the damage was done. Resentment was pointless. Thinking about what should be done next was a more productive response. The urge for reparation soon arrived. The spirit of the Blitz was rekindled. It was time for clearing away the fallen and replanting the new. Owners of chainsaws and JCBs did a roaring trade, and in some places more damage than the storm itself. The media had their own solutions. The trees had fallen because they were too mature, 'geriatric' as they liked to call them, and the solution was a policy of growing young, healthy trees, keeping woods in a state of perpetual adolescence, so there would be no tumbled veterans to grieve over. At Goodwood, close to where Alex Watt had watched his wild beechlings begin their slow struggle towards adulthood, I watched bulldozers scraping woods clear not just of fallen timber but of every living tree too, and replanting neat rows of identical nursery-grown seedlings. Plantable saplings, regardless of quality or species, became seen as a kind of panacea, as opportunist nurserymen and local politicians pressed for fast, publicity-grabbing action. The trees that were cleared were regarded as thoroughly corrupted, and went mostly for firewood. The furniture makers Habitat and Ercol considered launching a special brand labelled 'Saved from the Storm' but nothing came of it, despite the fact that 8 out of every 20 tons of fallen wood was beech. Fallen beech, they insisted, was unpredictable, often having internal flaws and shatter-zones which were invisible until it was planked. Furthermore it didn't meet the uniform standards their factories were used to receiving from plantations. Still, enough fallen beech was harvested to glut the British market, and 1,500 tons from the National Trust's estate at Slindon in Sussex was shipped to Turkey, where less fastidious joiners turned it into tables and chairs.

And then suddenly it was spring, and things began to look different. As people ventured timidly out into their transformed woodlands they began to notice that many of the trees presumed to be dead were putting out sheaves of leaves, and the ground around them was covered with millions of released seedlings. In plantations where large numbers of insubstantial trees had simply been sucked out or snapped off, there was an unprecedented show of spring flowers in the gaps. When I went back to Noar Hill in Hampshire it looked like an energetic woodland rehabilitation ward, a workout by prosthetic trees. Ashes which had been stripped of their branches by collapsing neighbours were wrapped in muffs of epicormic shoots. Fallen oaks had resumed activity in the horizontal mode, and were sprouting like newly laid hedges. I remembered the fable quoted by Gabriel García Márquez in *One Hundred Years of Solitude*, of the sailing ship beached in the rainforest, its masts breaking into leaf. All over the woodland floor were the little pinnate leaflets of seedling ash, relishing their chance in the sun. Even some of the big fallen beeches were coming into leaf, half their root-systems having overwintered in sodden earth. On one root-plate I found a clump of primroses blooming 2 metres above the ground, and old-man's-beard cascading down into the pit. The Hanging Gardens of Hampshire.

Woodland ecologists had hoped this was what would happen, that it would be a great opening-up of a wooded landscape that had perhaps become too dark, too homogeneous, too prone to the perils of uniformity. Storms of this severity are rare measured by the scale of human lifetimes, but regular enough in the long evolution of trees to have been incorporated in their genetic memories and recovery modes, in natural woodland at least.

Down in the National Trust's woods at Toy's Hill in Kent, probably the most comprehensively devastated site in the county, ecologists were already at work, agog at this opportunity to see what happened to a wood hit by a full-scale natural catastrophe. David Hutton took on the precarious task of surveying one of the woods amongst a head-high lattice of fallen trees, and produced a map of a strange and brittle beauty. It echoes Joseph John's map of the standing and fallen trees in Boubinsky Prales, but was generated with computer graphics. The forked and pollard beeches lie about like tangled tuning forks, in amorphous pools of hatches and dots which mark the overlapping mats of holly scrub and canopy debris. It's a diagram of dissonance, but a picture of character too, catching the roughneck quality of wild English beech pollards, compared to the austere columns dying more gently of old age in the forests of 19th-century Czechoslovakia.

When I visited Toy's Hill myself, the National Trust was at work, too. Their foresters, rather more optimistic than their exhibition catalogue writers, had sawn off some of the beech trunks, and pushed the root-plates back in their holes. One had created a splendidly rude piece of chainsaw sculpture with a forked log and erected it like a fertility symbol, standing out over the new views of the Kentish Weald. Hoping perhaps to conjure new trees into life. In fact the ground was already covered with seedling birch, beech, ash and yew. But a few hundred yards away, the Trust's workers were attempting to salvage fallen trunks and clear away the debris with bulldozers, in the belief that the 'public perception of natural woodland did not include fallen and dead trees'. Some of the patches of gravel they scraped bare were still without vegetation five years later. By contrast, in areas where fallen trees had been left on the ground, seedling beeches were already pushing through the wreckage, in accordance with their ancient preference for starting life in shady and protected nooks.

The National Trust took note. When a second major storm hit southern England in January 1990, leaving it with more devastated properties than it had the resources to restore in conventional ways, it commissioned a review of its whole tree and woodland management policy. The author was David Russell, the Trust's new forestry adviser, and a man with adventurous views about the importance of wildness to the human spirit. (After his term at the Trust, he became a pyschotherapist: such resonances between the forest and the mind!). In careful and moderate language, he reorientated the forestry-based principles on which the Trust's 25,000 hectares of woodland had previously been managed. From now on it must respect 'the way in which woodland operates as a natural system, and the vulnerability of that system to sometimes routine silvicultural practices'. Woods were 'better able to sustain themselves than has always been appreciated', and the observation of their doing this was an education, for public and professional alike. As the largest conservation charity in the country, the Trust had not always been sensitive in its alliances. It had made 'too great an investment in timber to the detriment of conservation', been too tolerant of the chemical farming practices of its tenants on wooded estates, and too easily carried away by the contemporary passion for planting. 'Greater care is needed in the choice of planting site than has always been appreciated.'

And so, on the Trust's properties at Ashridge and Frithsden, including the hanging beechwoods at Ivinghoe where I'd gawped at toadstools twenty years before and Paul Nash had painted his geological (or trichological) study of *Wood on the Downs,* the skittled trees were mostly allowed to remain where they'd fallen. Those blocking footpaths or potentially dangerous were logged up, but the remainder were allowed to play out their own destinies. There was no clearing-up of debris, no replanting.

I explored the hangers many times over the years that followed, watching the regenerating wood begin to take shape. And I walked the Chiltern scarp one summer's day fifteen years on from the 1990 storm, all the way from the Queen Beech to the open downland at Ivinghoe. The Beeches themselves had barely been touched by the storm, though the birches that were developing on the once open common had been raked through. It was the south-west-facing side of the escarpment that had been hit by the full force of the 1990 tempest. Whole blocks of forest beeches had been flattened. Now it was a different kind of wood. Sheaves of ashes had shot up between the fallen trunks, and were now 5 or 6 metres tall. The trunks themselves were slowly rotting away, but the root-plates were still jutting into the air. They'd become miniature ecosystems in their own right, homes for wrens and solitary wasps, cloaked with dog-rose and wild strawberries. (In some woods in Sussex, root-plates were nested in by kingfishers.) The ground around them had a quite new topography, a chaos of pits and 'tip-up mounds' which had added a huge diversity to the micro-habitats of the wood. They had also brought mineral-rich soils up to the surface, a great boon for regenerating seedlings. There were a few beeches amongst the ash, but their best display was on the edge of the open downland. They were shooting up – some already 2 or 3 metres high – out of the grassland in the lea of the wood, a place that now looks on the cusp of wood and down, an intermingling of tongues of grassland and wedges of saplings. A 'debatable' landscape. It was not exactly how they should have been regenerating, but such fixed ideas are now themselves part of history.

The storms of 1987 and 1990 had an impact far beyond the treescapes of southern England. They brought down, probably permanently, the idea that woods were settled places, whose behaviour could be tidily predicted in human models. They showed that natural disturbances were an entirely normal and well-tolerated part of a woodland's experience. It prompted changes in conservation policy, too. It helped contribute to a major shift in the National Trust's attitude towards their landscapes, and to the reforms of the Forestry Commission, which now takes a much more positive view of the natural elements of its wooded estate. Its furthest ripples have even reached the Wildlife Trusts, which have realised that their traditional policy of defending small static reserves – natural ghettoes, their own 'phantasy' retreats from the real world – is a dead end, and have begun moving increasingly towards the restoration and creation of big landscapes that need the minimum of human interference. But perhaps the biggest effect was on public opinion. The storms rekindled our affections for trees, and if there was perhaps a little too much hasty planting in consequence, it was a small price to pay for the restoration of an ancient partnership.

Chapter Eight: Vivat Regina

Global warming strikes the beech, but mysteriously it survives. Have we something to learn from its irrepressible variety, its adaptability, its quietness?

I

THE QUEEN BEECH survived both the 1987 and 1990 storms, and escaped, by a whisker, the great assault on ancient woods between the 1950s and '80s, that Oliver Rackham has called 'the Locust Years'. During not much more than twenty-five years, one-half of all England's ancient woodland was lost, as the Zeitgeist of the 'white heat of technology' penetrated the forest sector. It was an act of collective vandalism that has no parrallel in our landscape history, a massacre in which woods were devastated at a rate in excess of even the 18th- and 19th-century enclosures. Whole oakwoods were simply poisoned, or clear-felled for conifers. In 1967 Stanstead Great Wood, one of the biggest ancient woods in Suffolk, was sprayed by helicopter with 2,4,5-T, one of the components of Agent Orange, which the Americans used to deforest Vietnam. Beechwoods, often

growing on chalky soils and steep slopes, escaped comparatively lightly. But many woods in the Chilterns were cleared of their timber and replanted with mixtures of identical nursery-grown saplings and conifer nurse-trees. On the Ashridge estate, just a couple of hundred yards from Frithsden Beeches, the National Trust felled most of Frithsden Great Copse, and replanted it with conifers. It was a beautiful wood of hornbeam, maple and beech, surrounded by medieval banks, and was ravaged to commemorate Queen Elizabeth II's accession.

The coniferisation of these ancient woods has proved, in many places, to be a costly failure. The conifers – not of much value at the best of times – haven't always grown well. Many of the woods have been neglected or abandoned, and in some, the original trees have grown back from still-living stumps or seeds, and begun to overtake the conifers. In a few even the ground flora has begun to return, in areas not entirely sterilised by conifer needles. In their new, ecologically responsible clothes, both the National Trust and the Forestry Commission have policies of returning as many ancient woods to their original state as is practicable.

The Queen Beech seems also to have escaped the predicted effects of climate change. Even in the baking summers of 2003 and '06, it was hard to find older beeches showing symptoms of dehydration, as they had during the 1970s. And during the hot and early spring of 2007, where in Norfolk the oaks came into leaf unprecedentedly early in mid-April, the beeches stuck to their traditional date of late April – conserving water again maybe. Was it possible that the more drought-sensitive individuals had already been weeded out, and that those that remained had genetic qualities which gave them a measure of immunity to the heat?

*

I wanted to see how beeches in the extreme conditions of the south of Europe were coping, and Polly and I travelled down to the Mediterranean at the end of October 2006. It was soft warm autumn weather as we travelled through the Cévennes. The woods were tinged with the tans of oak and chestnut and beech. We were heading for the part of France I knew best, the limestone *causse* country around the River Dourbie, which I'd last visited nearly ten years before. I took Polly along the river, where I'd once watched beavers doing their own modest coppicing, and realised how many of the details of the local woods I'd missed then, entranced most of the day by nightingales and orchids. There were beeches throughout the riverside woods, growing amicably with oak and lime and service trees.

We walked north along the river gorge, with the hanging woods towering above us. Against the low autumn sun, they were dazzling. The ochre sea of oak was broken by rivulets of golden beech, flowing down the damp gullies and spring-lines of the hillside, down to the edge of the river, then *into* the river. Beeches with roots like mangroves were growing in the water. This was a riverine beechwood, a flood-plain forest in which this haunter of the uplands looked as much at home as the lemon-yellow poplars. I shouldn't pick favourites in a book like this. But these were the beeches that stole my heart, their silver trunks and bronzed leaves framed against the pale limestone cliffs and bubbling rapids, reaching their watery fulfilment at last.

I wasn't quite sure when we reached the true south. We'd turned east again, and were staying at a bed-and-breakfast near Valence, sitting on beechwood chairs made by the owner, who had turned out to be a joiner. I asked his wife how close we were to the Mediterranean, and got into a muddle trying to find words for what I really meant, which was the Mediterranean bio-region. But I got there in the end, by asking about lavender. It began just a little further south, she laughed:

their village was 'Midi moins quart'. And, sure enough, 10 kilometres on, it became Midi complet. The sun came out. The air smelt of resin. The side roads were draped with cistus and rosemary, just beginning its second flowering. We picnicked on a hillside amongst the juniper and lavender, looking out over a Provençal landscape in which rock and garrigue and wood seemed to have no seams. But the beech had vanished, up into the cool of the hills.

Mont Ventoux is rising just to our east. We drive up the warmer south side, through tracts of oak and peach-pink Montpellier maple, then immense plantations of Austrian black pine. The beech begins at about 750 metres, just a few seedlings amongst the pines at first, then whole tracts of high forest. At 1,500 metres it vanishes along with most of the other trees. The very top of Mont Ventoux, at 1,740 metres, is an awesome place, a desert of bare limestone clitter raked by winds of up to 180 mph. The Italian poet Petrarch, the first literary mountaineer, stood here in April 1336, amazed at a view that stretched from the Bay of Marseilles in the south to the mountains around Lyons, but not able to restrain himself from pious moralisings. 'How earnestly we should strive', he wrote, 'not to stand on mountain-tops, but to trample beneath us those appetites which spring from earthly impulses.'

We drop out of the wind, and begin walking down the mountain, full of earthly appetites. We find the wild beech about 150 metres below, tucked into little sheltered coombes and gullies. They're not like any beech I've seen before. There are immense coppice-stools, beeches growing out of limestone scree, beech-tufts no bigger than lavender bushes. The biggest are shaped like low mushrooms, wind-pruned down to a height much less than their breadth. They are hard to measure. The long, serpentine lower branches are growing out of a woody circle about 15 metres in circumference, but the outer growth is muddled with saplings. In the benign lowlands of England a tree this immense might be 800 years old. Up in this relentless wind, growing on bare limestone, under snow for six months of the year, receiving only about 10 centimetres of rain in the summer, I can't even begin to guess at its age. Van Gogh, who knew the Provence beeches, had painted a scene like this not long before he died. The disorientating *Tree Roots and Trunks* (1890) is a tumult of grey-blue wood and flaying green, with no skyline and no groundbase, in which it's impossible to disentangle the root spars from the shooting stems.

Nor can I guess the histories of the smaller trees, the twiggy clumps, the coppice-looking bushes that are tilting out of unstable 45° slopes. I can see their roots snaking down the mountain, not burying themselves into the depths of the rock, but burrowing like reptiles just under the surface, water-seeking. I'd seen the same root formations flowing down from the 4,000-year-old cypresses on the White Mountains of Crete. Had all these smaller trees been cut by humans too? Or pruned by ice storms, furious gales, wandering sheep? It scarcely matters; any shepherd who can cut coppice while hanging onto a limestone scree qualifies as a natural force in my book.

And the trees are all producing fertile mast, contrary to French forestry wisdom that beeches are played out after about a century and a half. We can see lines and patches of saplings, growing in crevices and on miniature plateaux where the seed had been swept by rainwash or meltwater. Then Polly spots the leaf mound, a great pillow of beech leaves caught up in the prickly branches of a circle of juniper. I remember that the softness and amiable rustlings of beech leaves had made them a favourite mattress-filling for French peasants. They called them *lits de parlement*, 'talking beds'.

Circular juniper clumps, as symmetrical as fairy rings, line the path back up to the top of the mountain. They show no signs whatever of being invaded by beech, according to classic succession theory. Is this desolate site the true beech climax, a point of such extreme hardship that no greater disturbance is conceivable, and where these vegetational arrangements have lasted unchanged for millennia? I think of the immense genetic variety of this supposedly finicky species that has enabled the Mont Ventoux chapter to survive ice storms and gales and a summer rainfall of less than 15 centimetres, and take some of their miraculous mast back to England.

II

I'm taking a different path to the Queen this time, tacking along the edge of the golf course that was created on the common in the 1920s, moving back in time through the heather rough and the Beeches' fringe of oak and birch. The track is strewn with auguries. A lost golf ball caught in a chaplet of bluebells. Bits of a deer's leg, brought up from the road by a fox. And at the bottom of one of the oaks a wreath of plastic leaves and berries, a funerary gift. There's a grey powder heaped over the oak's roots, a slight odour of scorch. Someone who loved the Beeches had their ashes scattered here, and they picked on the wrong tree.

I drift into the core of old pollards, past familiar friends. The Praying Beech, shrunk to a crumbling grave-mound the colour of peat. Falstaff, still standing, with its magnificent paunch. It looks like a tree slowly melting, slumped into its lower reaches. But insidiously the whole place begins to look unfamiliar. I'm walking on a lawn the size of a tennis-court, except that the turf is beech seedlings, deer-browsed down to a few inches in height. It's like bouncing on a trampoline. The beechlings are springy, spiky. I wonder what they feel like to a cloven hoof. The deer have munched their way so far, but seem to have given up where the growth is thickest. Ragged saplings, up to a metre tall, are poking clear of the twiggery, and receiving no more than a perfunctory nibble. It looks like an exercise in amateur topiary, a beech hedge dragged through itself backwards.

I'm drawn towards the fallen tree that opened this gap and has given the seedlings their chance in the sun. Its huge root-plate is a wooden cliff, a vertical quarry. Flints are bedded tight and close between the root-fibres like shoals of fossils. Even after days of rain the sandy earth around them turns to powder when I touch it. Rearing into the sky, its roots desiccated, it looks like an emblem of the whole catastrophe of the beech. Beech as victim. Precarious, vulnerable, parched into oblivion. But half its roots are still in the ground, and the trunk has sprouted a new pole, 15 metres tall, and contorted to a barley-sugar stick by squirrels.

Squirrels. Without meaning to I've drifted down into their domain, a narrow valley in the wood that stretches back to the road. They like working concentrated areas, the convenience of clustered buffets. Down here they've chewed great patches of bark off low trunks, ring-barked small trees, stripped whole crowns, but barely touched a group of older beeches 50 metres away. They're not behaving at all like the squirrels in Lady Park Wood. The trees they've attacked are mostly about 10 to 20 years old, with diameters no more than 10 centimetres. They look wretched, poles of scar tissue rather than orthodox trees. Round the bare patches of exposed heartwood the bark has formed long twisted lips of thickened repair tissue. The beech's stiff upper lip. Sometimes the squirrels gnaw this sappy stuff too. The trunks are bubbling with scars upon scars, like a too-often-mended bicycle tyre. But not many have died yet. Looked at more generously they're dwarfed pollards. If any survive it's hard to imagine what they might become a century or so hence, grown out of their squirrel-lopped youth to the age of the beeches at Burnham.

I try to get back to the thin track that winds towards the Queen, but swing too far the other way, into the open common. I can't understand why I'm not finding the trail. The place has changed since I was last here, maybe three years ago. In that short interval the Queen has become a star, adding an 'atmosphere of menace' to movies. In Harry Potter films it's played the part of the infamous Whomping Willow, a mystical tree that, beech-like, has snaking branches that hit back when it's attacked. But no trace of the film crew's presence remains. My old landmarks seem to have shifted. The gaps have bushed up, the paths look unused and overgrown. Maybe people are put off by the warning signs. No hunger for a pollard on the head. And I sense I'm walking too free-form, going in circles round fallen trees, entranced by the pattern of saplings coming through. I'm trying to get a feel of how the whole wood is vibrating, that long, low frequency of succession. I want to see the 'beech wave', and it's swinging me out of my knowledge. For an anxious moment, I wonder if I've got a false memory, and the Queen has already fallen, been tidied away.

I go back to the centre of the wood and do a transect at right angles to the direction I've been taking, and at last I find the old track, almost sealed off by holly and bramble. I'm a couple of hundred yards along it when the Queen drifts into view, a great inverted bowl hulking through the mist of foliage. It still takes my breath away, the mass of it, the hunched shoulders, the low spreading skirt. I measure it with my arms – a little gesture of reverence, though I know the answer by heart: 10 metres around at chest height, about double that just a metre higher. Four great branches almost horizontal. An arching dome, the cranium of the wood.

I follow the roots away from the trunk, 12 metres, rippling out almost as far as the branches. One great branch has been lost, but the tree looks stable, as so much of its weight is down near the bottom. I walk backwards away from it, into the future. There are small hollies in the deep shade. Beyond them a few beechlings, waiting for their moment. The beech tideline has reached 100 metres beyond the edge of the wood, into the open birch and oak scrub – which is advancing the other way, *into* the Beeches. None of the young beeches is more than 2 metres tall. They aren't much more than just-opened parasols, but they are already being nibbled by deer and squirrels. I wonder what will become of them. Will they ever grow into what we think of as 'proper' trees, great columns that will eventually overtop and shade out the oaks? Or will the grand theories of succession be overturned, and the trees become a new kind of undergrowth, beech bushes, just as chronically diseased elms have become in hedgerows?

I imagine myself on the top of the Queen, like one of Miroslav Holub's myopic moles. I can't see any of the architecture of trunk and picturesque branch that we dote on from below. This high canopy is the focus of the beech's own life. Those battles by the seedlings against voles and pigeons, the endurings of drought and frost, the persistent defiance of gravity in the search for the sun, the sinuous, expressive repairs by the branches in the face of wild gales and human lopping – all that dogged grace under fire – are devices to give just one seed the chance to succeed. From this high point I could see across a thousand years of beech imaging, a potted history of our attempts to put trees' self-interested determination inside some kind of human-serving frame. A few hundred yards to the south is a small plot where conservationists have tried to stop the advance of the oak and restore the heather, which had been here for maybe two thousand years before grazing was stopped in the 1920s. But we were here first, say the oak and beech. A mile further on a new by-pass is lined with planted and self-sown beeches, doomed to be regarded as a safety hazard in a hundred years' time. Then Hardings Wood on the top of the next ridge, in much more secure ownership now, a village trust constituted to protect it in perpetuity. Will they allow the plantation beeches to grow into *dépérissement*, like the trees in Fontainebleau? Will climate change get them first? If so what will the wood become then?

The same conundrum faces the trees immediately below my high vantage point. What will happen to the Beeches? The old trees are threatened by drought and gales and fungal invasions, the young by squirrels and deer. Should we intervene? Slaughter the squirrels, cull the deer, cut out every tree that might conceivably be vulnerable to wind? Replant and protect? Or abandon beech altogether, slaughter the dotards as dangerously unsuitable for local conditions and insidious warming, opt for the reliable oak?

My own view is that, instead, we should give the beeches a chance to sort things out for themselves. The lesson of the historical and geographical journey I've taken through the beech nation is that the tree has powers of survival we've scarcely begun to understand. The dwarf and drought-resistant veterans of wind-racked, snow-covered Mont Ventoux and the rheumatic survivors on Mount Pindos are the stock from which our northern beeches sprung, and our English trees have their genes somewhere in their own populations. For me, allowing them their chance is a moral issue. The idea – still argued by some conservationists – that all woods must be managed is as arrogant and outrageous as suggesting that all wild animals should be in zoos.

Managed woods reflect too simplistically our own limited skills and horizons. Wild, unmanaged, trees show us possibilities beyond our cultural tunnel-vision. They are ravishing in their autonomous lives and mercurial beauty. Their ingenuity and adaptability teach us lessons about our limitations and presumptions. They are greater than any of the images or roles we can confine them in. And part of this magnaminous flexibility may be a different kind of 'usefulness' from the human-centred utility we usually seek from the natural world. Trees have evolved through aeons of climate change. Collectively they know how to cope with it. We don't, and need to learn from solutions that they may only be able to express in unmanaged and unmanipulated situations. Their ancient, inbuilt diversity is not available from nursery-grown stock.

Wild trees and natural woods, untouched by (though not unentered by) humans are, I believe, essential to the planet's survival. But we live in the world, too, and need trees as raw materials and ornaments to our everyday lives. In America they use the phrase 'the middle landscape' for this conceptual area that lies between the wholly wild and the wholly cultivated. Can trees teach us something about how to live in this place without the overarching dominance whose history I've traced in this book?

Traditional practices, dynamic relationships such as coppicing and pollarding, are far from obsolete or irrelevant, provided they are not done as an unthinking historical reflex ('this is what has always been done'). They prolong the life and health of trees as well as producing benefits for us. It seems to me we have as much 'right' to work with trees as beavers and bark-beetles. Planting can be responsible and 'natural', too, provided it is done with the same sensitivity to soil and slope and patterning as happens spontaneously in the wild from the natural sowing and settling of seeds.

This book has been about the narratives of trees, the framings and images we make of them. It has also touched on the trees' own narratives, the stories they weave themselves in their migrations and distributions, in the groupings and patterns they form, even in their physical responses to being moulded by humans towards beauty in one direction and utility in the other.

But perhaps we need a spell not of stories, of monologues, but of conversations, mutual exchanges. I'm much taken with the idea of symbiosis, with that intimate, unshowy relationship between tree and fungus deep in the earth, and with that bonding between fungus and alga that forms a trunk-bound lichen. The relationship is essential to both partners, but it is carried on so that neither of them compromises the other's identity. Could we think our way back to a symbiosis with trees, where, in everything that we do with them, we consider what is good for the tree community *on its own terms* as well as what we can get out of it?

Afterword: Ashes to Ashes

It's the first winter after ash die-back was detected in Britain, and I'm back in the Chilterns. Frithsden Beeches seems much as it was a couple of years ago. No ashes to be afflicted by the fungus here. But many of the pollards have begun to sprout the formidable horses' hooves of *Ganoderma* bracket fungi, like the beeches in Fontainebleau. It will be the end of them eventually, but only after decades of interesting decay, accompanied by other fungal invasions and opportunist resproutings from undamaged greenwood. Meanwhile, spreading thickets of young beech and birch, some corkscrewed by squirrels, are invading the open patches.

On the opposite side of the valley, Hardings Wood looks unchanged, too, and there are no signs of Chalara yet. But I realise what has changed dramatically is my perspective on the place. With die-back on my mind I'm seeing ashes everywhere. I'm astonished that I could have visited this wood every week for the best part of twenty years and not fully taken in how ubiquitous they are. It was that pale, reticent presence, the smooth 50 year-old trunks rising without any side branches for 50 to 60 feet. In the summer, with their similarly unprepossessing fish-bone leaves, they're upstaged by the busy snaggings and broad leaves of the oaks and cherries. But in winter they become visible again. I reckon they must amount to between a quarter and a fifth of all the trees in the wood. If Chalara hits hard, and in the worst scenario, takes 95 per cent of the trees, that may leave less than a score of mature trees, which is much the same as the current scarcest species, sessile oak and rowan.

I tack up the path to the half-acre patch we clear felled in the late 1980s. The ashes have put on another 6 feet in the last three years, and there are fewer of them, as they out-compete each other. But there are still 13 other species of self-sprung shrubs and trees in this small community, so its future is secure.

But down in declivity near the eastern edge of the wood, where the soil is more calcareous, the damage is likely to be place-changing. Tall ashes make up maybe half the trees here. I move to the head of this tiny valley, look straight down it, and, in fast-forward, try to picture what may happen. The ashes infected, but not all at once, and dying slowly from the top down; the beech, oak and hornbeam spreading their branches as the ash canopy lets in more light; sycamore invading the open patches from its stronghold at the edge of the wood. In 20 years' time the canopy may have closed again, and there will be a matrix of more widely spaced trees interspersed by a few standing ash skeletons, and many fallen ones, breeding grounds for new fungi. The bluebells and wood anemones that are Hardings' great spring spectacle will be entirely unaffected.

It will, I suspect, be a little like a more than usually floral, underage version of a Chiltern wooded common. Not an ashwood province any more but still a living wood.

Acknowledgements

All book-writing turns the author's friends and acquaintances into a support group, a community of suggestion-makers, eyebrow-raisers, library-hunters, cuttings agencies, living encyclopedias, hosts, walking companions and hand-holders. My thanks to the following for so generously and creatively fulfilling these roles; Andrew Branson, Clive Chatters, Jeff Cloves, the late Roger Deakin, Bob Gibbons, Vivien Green, Francesca Greenoak, John Jackson and the staff at the Royal Forestry Society, John Kilpatrick, Robert Macfarlane, Peter Marren, the Norfolk Library Service and the staff at their branch in Diss, Oliver Rackham, Richard Reeves, Elizabeth Roy, Neil Sanderson and John Workman. And particular thanks to Susanna Wadeson, who in suggesting a rather different kind of book about trees, sparked off something I had been hoping to write for a decade.

George Peterken and Jonathan Spencer read various parts of the text, and I am indebted to them for pointing out some of my ghastly scientific solecisms.

Penny Hoare, my editor, was a model of patience and encouragement, as usual, and made truly imaginative suggestions about restructuring parts of the text. I'm continually indebted to her for her support. Jenny Overton, who copy-edited the book, did so with great diligence and professionalism, especially as it is couched in many different kinds of language.

Finally a mate's thanks to Polly, as always a source of down-to-earth wisdom, a sharp-eyed and witty companion on my research trips, and an understanding comforter during the bad patches.

References

Adams, Steven, *The Barbizon School and the Origins of Impressionism*, 1994

Addison, William, *Epping Forest: Its Literary and Historical Associations*, 1945

Bell, Thomas (ed.), *The Natural History and Antiquities of Selborne, by Gilbert White, including his correspondence*, 1877

Benjamin, Walter, 'The Work of Art in the Age of Mechanical Reproduction', in his *Illuminations*, 1970

Bowle, John, *John Evelyn and his World*, 1981

Broadmeadow, Mark, (ed.) *Climate Change: Impact on UK Forests*, Forestry Commission, 2002

Brown, David Blayney, 'The Nature of Our Looking', in David Dimbleby, *A Picture of Britain*, 2005

Burgess, Jacquelin, *Growing in Confidence: Understanding People's Perception of Urban Fringe Woodlands*, Countryside Commission, 1995

Burke, Edmund, *A Philosophical Enquiry into the Sublime and the Beautiful*, 1757

Buxton, E., *Epping Forest*, 1885, 5th edn 1898

Cameron, Laura and Forrester, John, 'A Nice Type of the English Scientist: Tansley and Freud', *History Workshop Journal*, 48 (1999)

Cardinal, Roger, *The Landscape Vision of Paul Nash*, 1989
Carpenter, Edward, *Pagan and Christian Creeds*, 1904
Clements, F. E., *Plant Succession and Plant Indicators*, 1916
Cobb, Rev. J., *The History and Antiquities of Berkhamsted*, 1855, new edn 1988
Cobbett, William, *The Woodlands*, 1825
Cole, Rex Vicat, *The Artistic Anatomy of Trees*, n.d.
Corona, Mauro, *Le Voci del Bosco*, Ponderone, 1998
Corke, David (ed.), *Epping Forest – the Natural Aspect?*, 1978
Countryside Commission, *The New Forest Landscape*, 1986
Daniels, Stephen, 'The Political Iconography of Woodland', in Denis Cosgrave and Stephen Daniels (eds), *The Iconography of Landscape*, 1988
Daniels, Stephen, *Fields of Vision: Landscape Imagery and National Identity in England and the USA*, 1993
Daniels, Stephen and Watkins, Charles, *The Picturesque Landscape: Visions of Georgian Herefordshire*, Nottingham, 1994
De Cleene, Marcel and Lejeune, Marie Claire, *Compendium of Symbolic and Ritual Plants in Europe*. Ghent, 1999–2003
Edlin, H. L., *Forestry and Woodland Life*, 1947
Edlin, H. L., *Woodland Crafts of Britain*, 1973
Evelyn, John, *Sylva*, 1st edn, 1664; 3rd edn 1679; 5th edn, ed. Alexander Hunter, 1776
Eversley, Lord, *Commons, Forests and Footpaths*, 1910
Fairbrother, Nan, *New Lives, New Landscapes*, 1970
Fisher, Roger, *Heart of Oak: The British Bulwark*, 1763
Foley, Michael and Clarke, Sidney, *Orchids of the British Isles*, 2005
Ford, Brian J., *Sensitive Souls: Senses and Communications in Plants, Animals and Microbes*, 1999
Forster, E. M., 'My Wood' (1926), in *Abinger Harvest*, ed. Elizabeth Heine, 1996
Forster, E. M., *Collected Stories*, 1947
Forster, E. M., *The Longest Journey*, 1984

Fowles, John, *The Tree*, 1979
Fowles, John, foreword to Elaine Franks, *The Undercliff*, 1989
Furbank, E. M., *E. M. Forster: A Life*, 1977–8
Gilbert, Oliver, *Lichens*, 2000
Gilpin, William, *Observations on the River Wye*, 1772
Gilpin, William, *Remarks on Forest Scenery*, 1791
Giono, Jean, *The Man Who Planted Trees*, 1985 (first published as an article in 1954)
Grahame, Kenneth, *Pagan Papers*, 1893
Grahame, Kenneth, *The Wind in the Willows*, 1908
Green, Peter, *Beyond the Wild Wood: The World of Kenneth Grahame*, 1982
Greene, Graham, *A Sort of Life*, 1971
Greig-Smith, P., 'A. S. Watt, FRS: A Biographical Note', in E. I. Newman (ed.), *The Plant Community as a Working Mechanism*, 1982
Grove, A. T., and Rackham, Oliver, *The Nature of the Mediterranean, An Ecological History*, 2001
Hadfield, Miles, *British Trees*, 1957
FL Harding, P. T., and Rose F., *Pasture Woodlands in Lowland Britain*, ITE, 1986
Harrison, Robert Pogue, *Forests: The Shadow of Civilisation*, University of Chicago Press, 1992
Hayman, Richard, *Trees, Woodland and Western Civilisation*, 2003
Heath, Francis George, *Burnham Beeches*, 1879
Hepple, Leslie W. and Doggett, Alison M., *The Chilterns*, 1992
Hervet, Jean-Pierre, and Mérienne, Patrick, *Fontainebleu: Une Forêt de Légendes et Mystères*, Éditions Onest-France, 2004
Hill, George, *Hurricane Force*, 1988
Hussey, Christopher, *The Picturesque: Studies in a Point of View*, 1923
Innes, J. L., 'Observations on the Condition of Beech in Britain, in 1990', *Forestry*, 65 (1992)
Johnson, Hugh, *The International Book of Trees*, 1973

Jones, Gather Lovett and Mabey, Richard, *The Wildwood*, 1973
Ketton-Cremer, R. W., *Felbrigg: The Story of a House*, 1962
King, James, *The Interior Landscapes of Paul Nash*, 1987
Kirby, Keith J., and Watkins, Charles (eds), *The Ecological History of European Forests*, 1998
Knight, Richard Payne, *The Landscape*, 1795
Land Use Consultants, *The Chilterns Landscape*, Countryside Commission, 1992
Le Sueur, A. D. C., *A Guide to Burnham Beeches*, n.d.
Linnard, William, *Welsh Woods and Forests*, National Museum of Wales, 1982
Logan, William Bryant, *Oak: The Frame of Civilisation*, 2005
Loudon, J. C., *Arboretum et Fruticetum Brittanicum*, 1844
Lowood, Henry, 'The Calculating Forester: Quantification, Cameral Science, and the Emergence of Scientific Forestry Management in Germany', in *The Quantifying Spirit of the Eighteenth Century*, University of California Press, 1991
Mabbett, Terry, 'Beech Bears the Brunt', *Essential ARB*, 14 (2004)
Mabey, Richard, *Home Country*, 1990
Mabey, Richard, *Flora Britannica*, 1996
Mansfield, A. J., 'The Historical Geography of the Woodland of the South Chilterns', University of London MSc thesis, 1952
Marren, Peter, *The Wild Woods*, Nature Conservancy Council, 1992
Massingham, H. J., *Through the Wilderness*, 1935
Massingham, H. J., *England and the Farmer*, 1941
Massingham, H. J., *Chiltern Country*, 1949
Mattheck, Claus, *Stupsi Explains the Tree: A hedgehog Teaches the Body Language of Trees*, Karlsruhe, 1999
Mattheck, Claus, *Design for Nature*, Springer Verlag, 1999
Matthews, J. D., 'The Influence of Weather on the Frequency of Beech Mast Years in England', *Forestry*, 28 (1955)

Meiggs, Russell, *Trees and Timber in the Ancient Mediterranean World,* 1982
Morris, M. G. and Perring, F. H., *The British Oak: Its History and Natural History,* BSBI 1974
Mountford, E. P., 'A Decade of Grey Squirrel Bark-stripping Damage to Beech in Lady Park Wood', *Forestry,* 70 (1997)
Nash, Paul, *Monster Field,* 1946
Ogley, Bob, *In the Wake of the Hurricane,* 1988
Orwell, George, *Keep the Aspidistra Flying,* 1936
Pakenham, Thomas, *Meetings With Remarkable Trees,* 1996
Pennington, Winifred, *The History of British Vegetation,* 1969
Peterken, George, *Woodland Conservation and Management,* 1993
Peterken, George, *Natural Woodland,* 1996
Peterken G. F. and Jones E. W., 'Forty years of change in Lady Park Wood', Journal of Ecology, 75 (1987) and 77 (1989)
Peterken G. F. and Mountford E. P., 'Lady Park Wood: the First Fifty Years', *British Wildlife,* 6 (1995)
Peterken, G. F. and Mountford, E. P., 'Effect of Drought on Beech in Lady Park Wood, an Unmanaged Deciduous Woodland', *Forestry,* 69 (1996)
Peterken, G. F. and Mountford, E. P., '60 Years of Trying at Lady Park Wood', *British Wildlife* (2005)
Peterken G. F. and Tubbs, C. R., 'Woodland Regeneration in the New Forest, Hampshire, since 1650, *Journ. App. Ecol.,* 2 (1968)
Porley, Ron and Hodgetts, Nick, *Mosses and Liverworts,* 2005
Postle, Martin, *Thomas Gainsborough,* 2002
Price, Uvedale, *Essay on the Picturesque,* 1794–8, in *Sir Uvedale Price on the Picturesque,* ed. Sir Thomas Dick Lauder, 1810
Quammen, David, 'Reaction Wood', in *Wild Thoughts from Wild Places,* New York, 1998
Rackham, Oliver, *Trees and Woodland in the British Landscape,* 1976
Rackham, Oliver, *Ancient Woodland,* 1980

Rackham, Oliver, *The History of the Countryside*, 1986
Rackham, Oliver, *The Last Forest*, 1989
Rackham, Oliver, *Woodlands*, 2006
Ratcliffe, Derek (ed.), *A Nature Conservation Review*, 1977
Reed, J. L., *Forests of France*, 1954
Roden, D., 'Woodland and Its Management in the Medieval Chilterns', *Forestry*, 41 (1968)
Rodwell, J. S. (ed.), *British Plant Communities*, 1997
Russell, David, *A Review of the Management of National Trust Trees and Woodland*, 1992
Schama, Simon, *Landscape and Memory*, 1995
Schama, Simon, *Power of Art*, 2006
Sebald, W. G., *Campo Santo*, 2005
Silvertown, Jonathan, *Demons in Eden*, 2006
Sinclair, Iain, *Edge of the Orison*, 2005
Spencer, Jonathan and Feest, Alan (eds), *The Rehabilitation of Storm Damaged Woods*, University of Bristol, 1994
Steuart, Sir Henry, *The Planter's Guide*, 1827
Stones, Jon and Rodger, Donald, *The Heritage Trees of Britain and Northern Ireland*, 2004
Strutt, Jacob George, *Sylva Britannica*, 1822. Folio edition, 1830
Summerhayes, V., *Wild Orchids of Britain*, 2nd edn 1968
Tansley, Arthur, *The New Psychology and its Relation to Life*, 1920
Tansley, Arthur, *The British Islands and Their Vegetation*, 1939
Taplin, Kim, *Tongues in Trees*, 1985
Thomas, Keith, *Man and the Natural World*, 1983
Townsend Warner, Sylvia, *Lolly Willowes*, 1926
Tubbs, Colin R., *The New Forest: An Ecological History*, 1968
Tubbs, Colin R., *The New Forest*, 1986
Tudge, Colin, *The Secret Life of Trees*, 2005
Vancouver, Charles, *General View of the Agriculture of Hants*, 1813

Vera, F. W. M., *Grazing Ecology and Forest History*, CABI, 2000
Wallace, David Rains, *Bulow Hammock: Mind in a Forest*, 1988
Warner, Marina, 'Signs of the Fifth Element', in *The Tree of Life*, South Bank Centre, 1989
Watt, A. S., 'On the Ecology of British Beechwoods', *Journal of Ecology*, 11 (1923), 12 (1924), 13 (1925)
Watt, A. S., 'Preliminary Observations on Scottish Beechwoods', *Journal of Ecology*, 19 (1934)
Watt, A. S., 'The Vegetation of the Chiltern Hills', *Journal of Ecology*, 22 (1934)
Watt, A. S., 'Pattern and Process in the Plant Community', *Journal of Ecology*, 35 (1947)
Whitbread, A. M., *When the Wind Blew: Life in our Woods After the Great Storm of 1987*, RSNC, 1991
White, Gilbert, *The Natural History of Selborne*, 1793; see also Bell
Whybrow, George H., *The History of Berkhamsted Common*, n.d. [*c.*1925]
Wildlife Trusts, *Living Landscapes*, 2006
Wilson, Edward O., *Biophilia*, 1984
Wiener, Yvette, Introduction to Mimi Lipton and Thorsten Düser, *Stacking Wood*, 1993

Index